Lecture Notes in Mechanical Engineering

Lecture Notes in Mechanical Engineering (LNME) publishes the latest developments in Mechanical Engineering—quickly, informally and with high quality. Original research or contributions reported in proceedings and post-proceedings represents the core of LNME. Volumes published in LNME embrace all aspects, subfields and new challenges of mechanical engineering.

To submit a proposal or request further information, please contact the Springer Editor of your location:

Europe, USA, Africa: Leontina Di Cecco at Leontina.dicecco@springer.com
China: Ella Zhang at ella.zhang@cn.springernature.com
India, Rest of Asia, Australia, New Zealand: Swati Meherishi at swati.meherishi@springer.com

Topics in the series include:

- Engineering Design
- Machinery and Machine Elements
- Mechanical Structures and Stress Analysis
- Automotive Engineering
- Engine Technology
- Aerospace Technology and Astronautics
- Nanotechnology and Microengineering
- Control, Robotics, Mechatronics
- MEMS
- Theoretical and Applied Mechanics
- Dynamical Systems, Control
- Fluid Mechanics
- Engineering Thermodynamics, Heat and Mass Transfer
- Manufacturing Engineering and Smart Manufacturing
- Precision Engineering, Instrumentation, Measurement
- Materials Engineering
- Tribology and Surface Technology

Indexed by SCOPUS, EI Compendex, and INSPEC.

All books published in the series are evaluated by Web of Science for the Conference Proceedings Citation Index (CPCI).

To submit a proposal for a monograph, please check our Springer Tracts in Mechanical Engineering at https://link.springer.com/bookseries/11693.

Subir Kumar Saha · K. Rama Krishna
Editors

Recent Developments in Multibody Dynamics

Proceedings of IMSD-ACMD 2020

Springer

Editors
Subir Kumar Saha
Department of Mechanical Engineering
Indian Institute of Technology Delhi
New Delhi, Delhi, India

K. Rama Krishna
Department of Mechanical Engineering
Indian Institute of Technology Delhi
New Delhi, Delhi, India

ISSN 2195-4356 ISSN 2195-4364 (electronic)
Lecture Notes in Mechanical Engineering
ISBN 978-981-97-7524-8 ISBN 978-981-97-7525-5 (eBook)
https://doi.org/10.1007/978-981-97-7525-5

This Springer imprint is published by the registered company Springer Nature Singapore Pte Ltd.
The registered company address is: 152 Beach Road, #21-01/04 Gateway East, Singapore 189721, Singapore

Preface

International Conference on Multibody System Dynamics (IMSD) is a biannual conference series, which started in 2010 with its inaugural edition in Lappeenranta, Finland. The following editions of the series took place in Lappeenranta, Finland (2010), Busan, Korea (2012), Stuttgart, Germany (2014), Montreal, Canada (2016), and Lisbon, Portugal (2018). The conference series serves as an opportunity for researchers, students, and professionals worldwide to exchange ideas in multibody dynamics concerning theoretical aspects, computational approaches, and in applied research such as vehicle dynamics, robotics, mechatronics, and biomechanics.

Asian Conference on Multibody Dynamics (ACMD), on the other hand, is another conference series for Asian countries. The first ACMD was held in Iwaki, Japan (2002). The following conferences were in Seoul, Korea (2004), Tokyo, Japan (2006), Jeju, Korea (2008), Kyoto, Japan (2010), Shanghai, China (2012), Busan, Korea (2014), Kanazawa, Japan (2016), and Xi'an, China (2018).

India was proud to host both the conferences for the first time in the country. The 6th Joint IMSD and the 10th ACMD conferences were originally planned to be held together in New Delhi in 2020. However, due to the COVID pandemic, they were deferred by two more years, and finally held in a hybrid mode during October 16–20, 2022. The conferences covered all the areas of multibody dynamics and attracted 185 abstracts, out of which 154 were accepted for the presentations in the conferences with an option to publish the full papers either in the Multibody System Dynamics journal or bring out as a separate publication. Finally, 92 papers were presented by the authors from several countries of the World, namely, from Asia, Europe, North America, and Oceania. Papers were presented under several themes: Biomechanics; Applications, Multidisciplinary Methods, and Other Topics; Benchmark Problems in System Dynamics; Computation Methods and Real-Time Applications; Contact, Impact, and Constraints; Control, Mechatronics, and Robotics; Dynamics of Vehicles; Flexible Multibody Systems; Modelling, Formalisms, and Theoretical Methods; Multibody Kinematics; MBD for Social Applications; and Optimization, Sensitivity Analysis, and Parameter Identification. There were more than 103 registrations, which included 50 physical (Foreign: 22 + Indian: 28) and 53 online (Foreign: 30 + Indian: 23).

However, the number of participants physically present was about 150 that included authors, presenters, keynote speakers, exhibitors, and volunteers.

Upon completion of the conferences, authors were communicated with the option of submitting full papers for peer reviews, to be considered for the Journal of Multibody System Dynamics (MSD) or the monograph to be published by the Springer who kindly agreed to publish selected papers. Based on the relevance, novelty, original contributions, and practical applicability, six papers were recommended to be published in the MSD Journal, and 16 papers were recommended for this monograph. We thank the authors for their contributions and the patience to see their papers published. The organization of a thorough review process could only be possible due to the efforts of expert reviewers. It would be unfair if we do not acknowledge the names of Mr. Ravi Tripathi and Mr. Alinjar Dan (Ph.D. students at IIT Delhi) who have been tirelessly coordinating with us in communicating with the authors and reviewers.

We also take this opportunity to thank the Indian Institute of Technology (IIT) Delhi for all its logistical support, along with several other institutes like IIT Jodhpur, IIT Roorkee, and Amrita Vishwa Vidyapeetham, Bengaluru, and a few others whose faculty and students contributed significantly to successfully plan overall several years, and organize the conferences in the midst of many uncertainties due to the COVID situations.

We thank the following societies and industries who have supported the conferences financially or otherwise:

Societies: IFToMM (International Federation for the Promotion of Mechanism and Machine Science); IUTAM (International Union of Theoretical and Applied Mechanics); AMM (Association for Machines and Mechanisms, India); The Robotics Society (India); ISME (Indian Society of Mechanical Engineers); KSME (The Korean Society of Mechanical Engineers); JSME (The Japan Society of Mechanical Engineers).
Industries: RecurDyn software by FunctionBay; DRDO; SOLIZE; HEXAGON; SVR INFOTECH.

Last but not least we thank Springer for considering our papers to be published in the form of a monograph.

New Delhi, India
April 2024

K. Rama Krishna
Subir Kumar Saha

Contents

Analytical Method for Determining the Collision-Free Workspace of an Over-Constrained Planar Cable-Driven 3D Printer 1
Ishan Chawla and Pushparaj Mani Pathak

Modeling, Simulation, Optimization of the DLR Scout Rover to Enable Extraterrestrial Cave Exploration 13
Antoine Pignède and Roy Lichtenheldt

Simulation and Control of Shape Memory Alloy Spring Actuator in a Flexible Tube Manipulator 33
Nisha Bhatt, Vedanshu Seedwan, Samyak Jain, Sanjeev Soni, and Ashish Singla

Design and Control of All Pneumatic Virtual Motion Simulator 43
Ashish Siddharth, Arun Dayal Udai, and Sourabh Khemka

Design of Flexure Robotic Hand for Teleoperation 65
Mohammad Zubair, Shubham Bhandari, and Indranath Chatterjee

Design and Development of a Novel Rotary Actuator Based on Shape Memory Alloy and Permanent Magnet System 73
Deep Singh, Rutupurna Choudhury, Yogesh Singh, and Santhakumar Mohan

UAV Landing on General Moving Platforms Without Markers 83
Sagar Dalai, Kanishk Vishwakarma, Kaushal Kishore, Dhruv Potdar, and Manan Arora

Innovative Two-Axle Vehicle with Improved Ride Comfort via Blended Active Vibration Control 93
Rocco Libero Giossi, Rickard Persson, and Sebastian Stichel

A Linear Frequency Domain Solver Workflow for Fast Simulation of Transmission Systems 105
K. D. Bauer, J. Haslinger, and G. Offner

Multibody Dynamic Study of Subassembly Transfer Flask Under Seismic Excitation 119
Sasidhar Inakollu, S. D. Sajish, and Jose Varghese

A Finite Element Analysis Study on the Effect of Tool Stiffness in Incremental Forming Process 139
Eldho Paul, Hariharan Krishnaswamy, Riby Abraham Boby, and Sahil Bharti

Mobile Haptic Device for Large Virtual Environments 149
P. Abinaya, K. S. Sasivarnan, Prasanna Kumar Routray, and M. Manivannan

Multibody Dynamics Software-Based Simulation of a Game for a Robotics Competition 171
Sandeep Kumar, Subir Kumar Saha, and Satinder Paul Singh

Modelling and Diagnosis of Defects in Spur Gear Under Constant Speed Operation 181
Rajeev Kumar, Ranjan Gouda, Samrat Mandal, and Chintamani Mishra

About the Editors

Subir Kumar Saha is a Professor in the Department of Mechanical Engineering, Indian Institute of Technology Delhi and Project Director of I-Hub Foundation for Cobotics (Technology Innovation Hub of IIT Delhi). He received his mechanical engineering degree from RE College (now NIT), Durgapur, India, in 1983, prior to completing his M.Tech. at the IIT Kharagpur, India, and Ph.D. at McGill University, Canada. After his studies, he joined Toshiba Corporation's R&D Center in Japan. After four years of work experience in Japan, he has been with IIT Delhi since 1996. He is actively engaged in teaching, research, and technology and has completed projects worth several crores of Rupees. He established the Mechatronics Laboratory at the IIT Delhi in 2001 and was Co-coordinator for the Programme for Autonomous Robotics at IIT Delhi during 2010–2016 sponsored by BARC/BRNS Mumbai, which lead to the formation of Centre of Excellence on Biologically Inspired Robots and Drones (BIRD). Recently, he was also instrumental to initiate M.Tech. in Robotics at IIT Delhi. In recognition of his international contributions, Prof. Saha was awarded a Humboldt Fellowship by the AvH Foundation, Germany, in 1999. He has been a visiting researcher in Canada, Australia, and Italy. Professor Saha has authored many research publications in reputed journals/conference proceedings and books, including one textbook on Introduction to Robotics supported with an in-house developed RoboAnalyzer software that has been licensed to a company in 2023.

K. Rama Krishna is an Assistant Professor in the Department of Mechanical Engineering at the Indian Institute of Technology Delhi. He received his Ph.D. from the Indian Institute of Science, Bengaluru. His interests are theoretical and applied kinematics, geometric modeling for CAD and manufacturing, dynamics and control of mechanical systems, differential geometry, and additive manufacturing of mechanisms.

Analytical Method for Determining the Collision-Free Workspace of an Over-Constrained Planar Cable-Driven 3D Printer

Ishan Chawla and Pushparaj Mani Pathak

Abstract Cable-driven parallel robots (CDPRs) have larger workspace, simple design, are easy to transport, assemble and disassemble, and hence, possess a great potential in on-site 3D printing of large structures. However, they are prone to collision of cables with the preceding layers of structures being 3D printed. Therefore, this work proposes an analytical approach to determine the boundary of the collision-free workspace of planar over-constrained cable robots. This workspace is composed of poses that can be reached while avoiding the collision of lower cables with the structure being printed. The analytical expressions of the collision-free workspace have been obtained for both cases, i.e., considering zero platform orientation and variable platform orientation. These analytical expressions have been utilized for rapid computation of collision-free workspace. Simulation results have been obtained for both cases, and the effect of orientation on the collision-free workspace is analyzed. Additionally, the optimal design problem of CDPR is presented to achieve the desired collision-free workspace while minimizing the installation space of the robot. The results show a significant reduction in installation space by considering the variable orientation of the platform.

Keywords Cable-driven · Over-constrained · Variable orientation · Collision-free workspace

I. Chawla · P. M. Pathak (✉)
Mechanical and Industrial Engineering Department, Indian Institute of Technology Roorkee, Roorkee 247667, India
e-mail: pushparaj.pathak@me.iitr.ac.in

I. Chawla
e-mail: ichawla@me.iitr.ac.in

S. K. Saha and K. Rama Krishna (eds.), *Recent Developments in Multibody Dynamics*, Lecture Notes in Mechanical Engineering,
https://doi.org/10.1007/978-981-97-7525-5_1

1 Introduction

Cable-driven parallel robots belong to the class of parallel robots in which rigid links are replaced by flexible cables. One end of the cables is attached to the end effector/mobile platform, while the other end is attached to the winches via pulleys. The end effector of the robot is manipulated by retracting and releasing cables with the application of winches in a fully automated way. Since cables are lighter in weight and can be extended to much larger lengths (with the application of a spool), CDPRs have the characteristic feature of a larger workspace and a higher payload-to-weight ratio. Based on these characteristics, CDPRs are often used in various applications, for example, in large-scale telescopes [1], motion simulators [2], video camera systems [3], logistics and 3D printing [4], etc. In this paper, the main focus is on the application of CDPR in 3D printing.

CDPRs can be classified as suspended CDPRs and over-constrained CDPRs. A CDPR is said to be a suspended CDPR if the mobile platform is supported by cables above the platform only, i.e., the robot entirely relies on gravity to be balanced. If the mobile platform of a CDPR is held by the cables both above and below the platform, then the robot is said to have an over-constrained configuration.

The literature on CDPRs for 3D printing can be categorized into two main parts: suspended CDPRs [4–7] and over-constrained CDPRs [8–10]. In suspended CDPRs, none of the cables is below the platform. Hence, these mechanisms are safe from encountering collisions with the object being printed. However, they possess low stiffness, which can lead to significant vibrations when subjected to external disturbances (such as hose pipe vibrations and wind) [11]. On the other hand, over-constrained CDPRs possess larger stiffness in comparison to suspended CDPRs. However, they encounter significant cable-structure collisions while 3D printing [11, 12]. Therefore, the collision of lower cables has to be considered while determining the workspace of these robots. As the application of additive construction requires high stiffness to external disturbances, the over-constrained redundant CDPR should be the most suitable configuration. Since stiffness is essential for on-site printing, this paper focuses on over-constrained CDPRs.

Determining the workspace of a CDPR is a challenging problem due to the unilateral constraints on the tension of cables. The wrench-feasible workspace (WFW) is obtained by finding a set of poses at which static equilibrium of the mobile platform can be achieved by bounding the cable tensions within the selected tension limits [13]. The choice of a positive lower tension limit is made to ensure the tangle-free operation of CDPR with controlled sagging. Alternatively, the selection of upper tension limit is done on the basis of the cable's breaking load and the actuator's maximum torque limitations, etc. [14].

Another major challenge is the determination of the collision-free workspace of the robot in which the printing can be done safely without encountering any cable-structure collisions. Thus, the printable workspace of the cable-driven 3D printer (CD3P) is the intersection of the wrench-feasible workspace with the collision-free workspace. A significant amount of work has been done in the literature on the determination of WFW and various methods are available in the literature for the rapid computation of WFW of the CDPR [15, 16]. Therefore, the focus of this paper is not on the computation of WFW but on the determination of the collision-free workspace of the CD3P.

The initial work on the printable workspace analysis of over-constrained CDPR is reported in [9], where the authors used an iterative approach to determine the collision-free workspace for bricklaying applications. The proposed methodology determines the printable workspace for the planar case by taking the projection of cables and nozzle on the vertical plane. This planar workspace is then extruded into the longer dimension of the frame to get the spatial printable workspace. Our previous work used this iterative approach to analyze the effect of geometrical parameters and optimally design the cable-driven 3D printer [17, 18]. In [19], an alternative approach based on optimization is utilized to find the largest printable parallelogram inside the collision-free workspace of a spatial CDPR. However, obtaining a collision-free workspace by using an iterative or optimization approach is computationally intensive, especially for the design analysis of CDPR. This is because, during design analysis, it is generally required to obtain a printable workspace for a large number of combinations of geometrical parameters to make an optimal selection of parameters.

Therefore, this paper proposes an analytical approach to determine the boundary of the collision-free workspace of a planar over-constrained CDPR. The analytical expressions of the collision-free workspace have been obtained for both cases, i.e., considering zero platform orientation and variable platform orientation. These analytical expressions have been utilized for rapid computation of collision-free workspace. Furthermore, the analytical expressions provide insight into the effect of geometrical parameters on the pattern of the printable workspace and can be significant in designing the robot for the desired printable workspace. Simulation results have been carried out for both cases and the effect of orientation on the collision-free workspace has been analyzed. Furthermore, the design optimization of the CD3P is carried out using a genetic algorithm and important observations have been highlighted.

The paper is structured as follows: In Sect. 2, the method for analytically determining the printable workspace is proposed for both cases, i.e., considering zero platform orientation and variable platform orientation. The simulation results and the corresponding discussions for the proposed approach are presented in Sect. 3. Finally, in Sect. 4, the conclusions are summarized.

2 Collision-Free Workspace

A four-cable planar CDPR is considered in which the mobile platform/nozzle is connected by cables in crossed configuration, as shown in Fig. 1. The exit points of the cable are located at corners of the rectangular frame of height h_b and length l_b. The mobile platform is characterized by geometric parameters such as nozzle tube height n_p, platform height h_p, and platform length l_p. The mobile platform of the robot can translate in x and z direction with respect to a global frame of reference

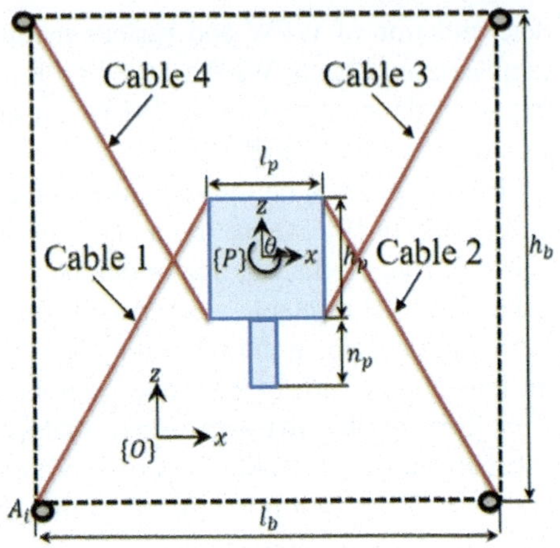

Fig. 1 Geometrical parameters of CD3P

$\{O\}$ and rotate in plane; thus, the robot has three degrees of freedom (DOF) (i.e., x, z, θ). On the other hand, the robot is actuated by four cables. Thus, the CDPR is redundant in nature as the number of cables is larger than the DOF of the robot.

In order to obtain the collision-free workspace, firstly, the line of symmetry l_s is identified to print the required object. The line of symmetry will be in the workspace center and parallel to z-axis in the case of no external wrench other than the platform's weight. A symmetrical collision-free workspace is obtained by maximizing the dimension of the object along the line of symmetry. For a symmetrical object and the cable arrangements, a cable-object collision can only happen between the lower cables and the object on the other side of the line of symmetry. For example, if the mobile platform is on the right side of the line of symmetry, cable 1 can collide with the structure on the left side of the l_s. Let x_b and z_b represent the x and z-coordinate of the boundary of printable workspace with respect to global coordinate frame $\{O\}$ located at the l_s, as shown in Fig. 2. When the mobile platform is at the right boundary of the workspace, cable 1 must collide with the object being 3D printed. Since the workspace is symmetrical, the point of collision P_2 will be at x_b distance from the line of symmetry. Using this knowledge, we can infer that cable 1 passes through three points, i.e., P_1, P_2 and P_3. The position of these three points with respect to global coordinate frame $\{O\}$ can be obtained in terms of geometrical parameters as follows:

$$P_1 = [-0.5l_b, 0]^T$$

$$P_2 = [-x_b, z_b]^T$$

$$P_3 = \left[x_b - 0.5l_p, z_b + h_p + n_p\right]^T$$

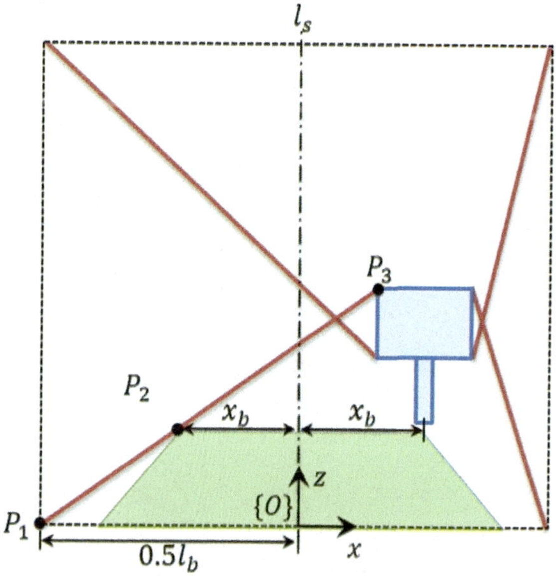

Fig. 2 Kinematics of cable-driven printing robot with zero mobile platform orientation

Assuming the cables to be of negligible mass, the cable follows a straight-line path joining all three points. The slope q of the line must satisfy Eq. (1).

$$q = \frac{z_2 - z_1}{x_2 - x_1} = \frac{z_3 - z_2}{x_3 - x_2} \tag{1}$$

Here, x_i and z_i represent the coordinates of point P_i (for $i = 1$–3) with respect to the global frame of reference.

By substituting values in Eq. (1) and solving, we can obtain the analytical expression for the boundary of the printable workspace as:

$$x_b = \frac{l_b(h_p + n_p) + z_b l_p}{4z_b + 2h_p + 2n_p}, \quad \text{where } z_b = 0 \text{ to } h_b \tag{2}$$

The printable workspace of the robot can be improved by allowing the mobile platform to change its orientation in order to prevent cable-structure collision. The change in orientation is carried across the tip of the nozzle and based on the orientation angle, the cable anchor point P_3 must lie on a circle of radius $r_c = \sqrt{\left(\frac{l_p}{2}\right)^2 + \left(h_p + n_p\right)^2}$. The chances of cable collision are minimum when the angle between cable 1 and the horizontal axis is maximum, which is only possible when cable 1 is tangent to the circle. For simplicity, consider a coordinate frame $\{Q\}$ located at the tip of the nozzle, as shown in Fig. 3. The equation of the line representing cable 1 and the circle in $\{Q\}$ can be given by Eq. (3) and Eq. (4), respectively.

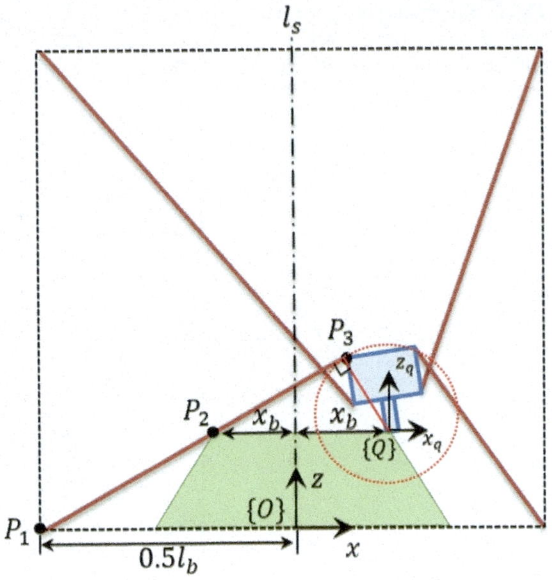

Fig. 3 Kinematics of cable-driven printing robot with variable mobile platform orientation

$$z = qx + c \tag{3}$$

$$x^2 + z^2 = r_c^2 \tag{4}$$

The point P_3 must lie on both the line and the circle. On substituting $z_3 = qx_3 + c$ in Eq. (4) and solving gives:

$$x_3^2 + (qx_3 + c)^2 = r_c^2$$

$$\Rightarrow x_3^2(1 + q^2) + 2qcx_3 + c^2 - r_c^2 = 0 \tag{5}$$

Equation (5) represents a quadratic equation in x_3. For point P_3 to be tangent, this quadratic equation must have a unique solution which is only possible when the following condition is satisfied.

$$(2qc)^2 - 4\left(1 + q^2\right)\left(c^2 - r_c^2\right) = 0$$

$$\Rightarrow q = \frac{\pm\sqrt{c^2 - r_c^2}}{r_c} \tag{6}$$

The slope q and intercept c can be obtained using point $P_1 = [-x_b - 0.5l_b, -z_b]^T$ and $P_2 = [-2x_b, 0]$ in $\{Q\}$ as follows:

$$q = \frac{z_b}{0.5l_b - x_b} \tag{7}$$

$$c = \frac{2x_b z_b}{0.5l_b - x_b} \tag{8}$$

On substituting q and c from Eq. (7) and Eq. (8), respectively, into Eq. (6) and solving for x_b gives:

$$x_b = \frac{-l_b r_c^2 \pm 2z_b r_c \sqrt{4z_b^2 + l_b^2 - r_c^2}}{8z_b^2 - 2r_c^2} \tag{9}$$

Since x_b should be positive, the $\pm$ operator in Eq. (9) can be replaced by $+$ operator. Thus, the analytical expression for the boundary of the printable workspace can be obtained as follows:

$$x_b = \frac{-l_b r_c^2 + 2z_b r_c \sqrt{4z_b^2 + l_b^2 - r_c^2}}{8z_b^2 - 2r_c^2}, \quad \text{where } z_b = 0 \text{ to } h_b \tag{10}$$

Thus, using Eqs. (2) and (10), the boundary of the collision-free workspace can be obtained for zero and variable mobile platform orientation.

3 Results and Discussion

The proposed approach is validated on a planar redundant over-constrained robot of the frame length $l_b = 10$ m and height $h_b = 10$ m. The workspace is obtained using MATLAB 2020a for two sets of platform geometric parameters, i.e., case 1: $[l_p, h_p, n_p] = [1, 1, 1]$ m and case 2: $[l_p, h_p, n_p] = [2, 2, 2]$ m, as shown in Fig. 4. The obtained results for case 1 show that the collision-free workspace of area 27.7 m^2 and 32.2 m^2 is obtained considering zero and variable orientation, respectively. Therefore, an increase in the collision-free workspace by 16.1% is obtained by considering a variable orientation. On the other hand, for case 2, an improved collision-free workspace of 42.2 m^2 and 52.3 m^2 area is obtained for zero and variable platform orientation, respectively. Therefore, a large increase of 23.8% is observed by considering variable orientation for large mobile platform.

The computational time for obtaining the workspace with the step size of 0.01 m is 0.1 ms for the proposed approach. In contrast, a high computation time of 254 s is required to obtain the collision-free workspace using the conventional iterative approach. Therefore, a significant reduction in computational time is observed using the proposed approach.

Furthermore, the design optimization problem of CD3P is addressed for the desired rectangular workspace of length 6 m and height 5 m. As the shape of the

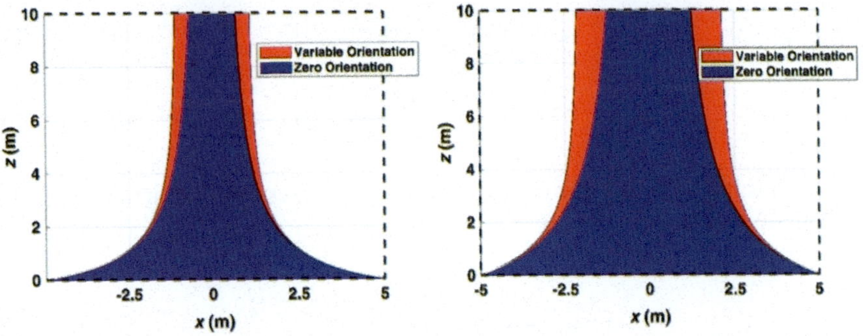

(a) (b)

Fig. 4 Collision-free workspace for the considered CD3P considering zero and variable orientation with the mobile platform parameters $[l_p, h_p, n_p]$ as **a** [1, 1, 1] m and **b** [2, 2, 2] m

collision-free workspace is somewhat similar to a normal distribution curve with mean at the line of symmetry, only the top corner (left/right) of the desired workspace can be checked to find whether the desired rectangular workspace lies within the printable workspace of the CD3P. This knowledge significantly reduces the computational time of the optimization problem as only one point has to be checked to know if the desired rectangular workspace lies within the printable workspace. Additionally, the number of variables in the optimization problem can be reduced by using the total nozzle height $h_n = h_p + n_p$ as a variable instead of platform height h_p and nozzle tube height n_p. In the design optimization problem, the objective is to minimize the installation space of the robot. Minimizing the installation space could result in a compact design of the robot. This could be beneficial in a situation where limited space is available around the structure to be 3D printed. Also, cost reduction could also be achieved since a smaller frame will be required. This can be done by minimizing the area of the frame (i.e., $l_b h_b$). Since the collision-free workspace is independent of frame height h_b, the performance index of the optimization problem can be simplified by just minimizing the frame length l_b. The optimization problem must also follow a constraint, i.e., the desired rectangular workspace must lie within the collision-free workspace of CD3P. It is intuitive that the increase in the mobile platform parameters leads to an increase in the collision-free workspace. However, this would also increase the weight of the platform. Therefore, in order to ensure that the weight of the platform is within the payload capacity of the CD3P, a constraint on the mobile platform parameters is added to the optimization problem. Thus, the cost function f of the optimization problem can be formulated as:

$$f(q) = \begin{cases} l_b \text{ if desired wokspace} \subseteq \text{collisionfree wokspace and } l_p + h_n \leq 6 \\ l_b + \text{HNV, otherwise} \end{cases} \quad (11)$$

Table 1 Parameters of genetic algorithm

Parameter	Value
Population size	600
Number of generations	2000
Selection policy	Stochastic uniform
Crossover rate	80%
Mutation rate	20%

Table 2 Optimal geometrical parameters

Parameters	l_b (m)	l_p (m)	h_n (m)
Optimized value (zero orientation)	16.45	0.5	5.5
Optimized value (variable orientation)	10.33	0.5	5.5

where $q = [l_b, l_p, h_n]$ is a vector of geometrical parameters to be optimized. The geometric parameters of the mobile platform are constrained by the minimum dimension of 0.5 m to prevent any parameter from converging at a significantly lower value. A genetic algorithm is utilized to solve this optimization problem. The parameters of the genetic algorithm are given in Table 1.

Since the proposed approach for boundary determination is analytical, the optimization converged in 6–8 s only. The results are obtained for both cases, i.e., zero and variable platform orientation, and the optimized geometrical parameters are given in Table 2. In both cases, the optimization converged at the lower bound (i.e., 0.5 m) for platform length and the upper bound (i.e., 5.5 m) for total nozzle height. This shows that the total nozzle height has more impact on the collision-free workspace in comparison to platform length. The optimization results showed that a minimum frame length of 16.45 m and 10.33 m is required to achieve the desired workspace for the case considering zero and variable platform orientation, respectively. Thus, a significant reduction of 37.2% in collision-free workspace is observed in frame length on considering variable platform orientation.

The workspace obtained using the optimized parameters for zero and variable platform orientation is shown in Fig. 5. The results show that the desired workspace (indicated by a dotted green line) is within the collision-free workspace of the CD3P.

4 Conclusions

In this work, an analytical approach is proposed to determine the boundary of a collision-free workspace for a planar cable-driven 3D printer. The analytical expressions for the boundary of the workspace are derived using geometrical relations for both zero and variable platform orientation. The simulation results are obtained for two different sets of geometrical parameters. Simulation results show a significant increase in the collision-free workspace on considering the variable orientation for

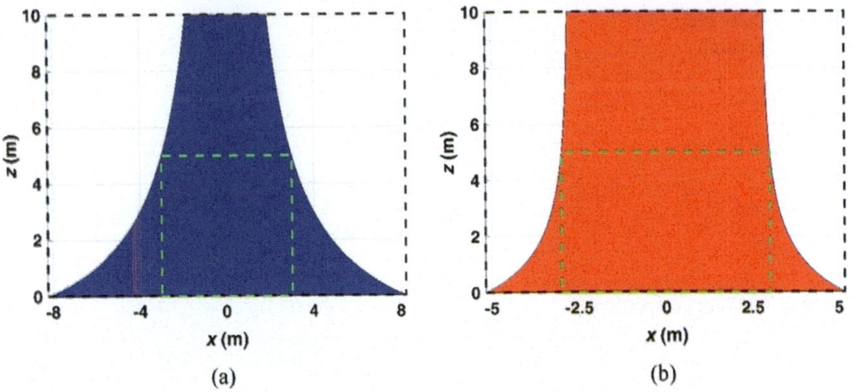

Fig. 5 Collision-free workspace for optimized geometrical parameters considering **a** zero platform orientation and **b** variable platform orientation

the platform. Furthermore, a significant reduction in computational time is observed using the proposed approach in comparison to the conventional iterative approach. The analytical expressions for the collision-free workspace give a clear insight into the effect of geometrical parameters on the shape and size of the workspace. Furthermore, the proposed approach is used to optimally design the CD3P by minimizing the installation space of the robot. A genetic algorithm is used for the optimization to achieve the global minima. The results show that the total nozzle height has more impact on the collision-free workspace in comparison to platform length. On the other hand, a significant reduction in the installation space is achieved by considering variable platform orientation in comparison to zero orientation. Therefore, the proposed approach is recommended for the design of cable-driven 3D printers.

The proposed approach can be extended to spatial robots in the future. One easy solution to this problem is to take the projection of the cable profiles onto the *xz* or *yz* planes, respectively, to simplify the collision detection. The obtained planar workspace can then be extruded into the longer dimension of the frame to get the spatial printable workspace in a similar way as reported in [9]. However, there is no guarantee that this solution will yield the largest possible printable workspace. Therefore, future work is recommended in formulating the analytical approach to determine the largest possible printable workspace.

References

1. Li D, Pan Z (2016) The five-hundred-meter aperture spherical radio telescope project. Radio Sci 51(7):1060–1064
2. Miermeister P, Lächele M, Boss R, Masone C, Schenk C, Tesch J, Kerger M, Teufel H, Pott A, Bülthoff HH (2016) The CableRobot simulator large scale motion platform based on cable robot technology. In: IEEE International conference on intelligent robots and systems, pp 3024–3029

3. Cone LL (1985) Skycam-an aerial robotic camera system. Byte 10(10):122
4. Izard J-B, Dubor A, Hervé P-E, Cabay E, Culla D, Rodriguez M, Barrado M (2017) Large-scale 3D printing with cable-driven parallel robots. Constr Robot 1:69–76
5. Barnett E, Gosselin C (2015) Large-scale 3D printing with a cable-suspended robot. Addit Manuf 7:27–44
6. Tho TP, Thinh NT (2021) Using a cable-driven parallel robot with applications in 3D concrete printing. Appl Sci 11(2):1–24
7. Dierichs K, Kyjánek O, Loučka M, Menges A (2019) Construction robotics for designed granular materials: in situ construction with designed granular materials at full architectural scale using a cable-driven parallel robot. Constr Robot 3(1):41–52
8. Bosscher P, Williams RL, Bryson LS, Castro-Lacouture D (2007) Cable-suspended robotic contour crafting system. Autom Constr 17(1):45–55
9. Wu Y, Cheng HH, Fingrut A, Crolla K, Yam Y, Lau D (2018) CU-brick cable-driven robot for automated construction of complex brick structures: from simulation to hardware realization. In: International conference on simulation, modeling, and programming for autonomous robots (SIMPAR), pp 166–173
10. Bruckmann T, Boumann R (2021) Simulation and optimization of automated masonry construction using cable robots. Adv Eng Inform 50:101388
11. Bruckmann T, Reichert C, Meik M (2017) Concept studies of automated construction using cable-driven parallel robots. In: 3rd International conference on cable-driven parallel robots, pp 364–375
12. Zi B, Wang N, Qian S, Bao K (2019) Design, stiffness analysis and experimental study of a cable-driven parallel 3D printer. Mech Mach Theory 132:207–222
13. Gouttefarde M, Merlet JP, Daney D (2007) Wrench-feasible workspace of parallel cable-driven mechanisms. In: International conference on robotics and automation, pp 1492–1497
14. Pott A (2014) On the limitations on the lower and upper tensions for cable-driven parallel robots. In: Advances in robot kinematics. Springer, pp 243–251
15. Guay F, Cardou P, Cruz-ruiz AL, Caro S (2014) Measuring how well a structure supports varying external wrenches. In: New advances in mechanisms, transmissions and applications. Springer, pp 385–392
16. Bouchard S, Gosselin C, Moore B (2010) On the ability of a cable-driven robot to generate a prescribed set of wrenches. J Mech Robot 2(1):011010
17. Chawla I, Pathak PM, Notash L, Samantaray AK, Li Q, Sharma UK (2021) Workspace analysis and design of large-scale cable-driven printing robot considering cable mass and mobile platform orientation. Mech Mach Theory 165:104426
18. Chawla I, Samantaray AK, Pathak PM, Li Q, Notash L, Sharma UK (2020) Miniaturization of a planar cable-driven 3D printer using optimization. In: Proceedings of MARSS 2020: international conference on manipulation, automation, and robotics at small scales, pp 6–11
19. Fabritius M, Martin C, Pott A (2018) Calculation of the collision-free printing workspace for fully-constrained cable-driven parallel robots. In: International design engineering technical conferences and computers and information in engineering conference, p V05BT07A046

Modeling, Simulation, Optimization of the DLR Scout Rover to Enable Extraterrestrial Cave Exploration

Antoine Pignède and Roy Lichtenheldt

Abstract In-situ exploration of extraterrestrial area of scientific interest remains largely impossible with the state-of-the-art wheeled rovers. To combine the advantages of wheeled and legged robots, DLR's Institute of System Dynamics and Control is developing a modular rover with rimless wheels, called *Scout*. To save time and money, the development of the Scout rover makes extensive usage of modeling and simulation. A detailed multibody model of the Scout rover accompanies the system through all phases following the V-shaped development process. This starts with the detailed modeling, analysis, and optimization of the rimless wheels and other central parts. It continues with simulation tests on the full system along with optimizations. For example, the properties of the "vertebrae" that connect segments together have been laid out following a digital optimization campaign. Subjects of this text are the details of the multibody Scout rover model, optimization on the rover and simulation analyses. A first prototype of the Scout rover has been built 1.5 years after the start of the project. It is already exceeding the requirements and expectations. Such good results would not have been possible this fast without simulation and optimization. Refinement of the simulation model and validation against the hardware are the next steps.

Keywords Modeling and simulation · Optimization · Cave and extreme terrain exploration · Extraterrestrial exploration · V-model development

1 Introduction

Current planetary exploration missions are based on robotic systems despite ongoing efforts for crewed missions back to the Moon and potentially forward to Mars. Robotic systems range from fly-bys and orbiters that remotely sense the target body, to landers and rovers for in-situ exploration. One notable example of an ongoing

A. Pignède (✉) · R. Lichtenheldt
Institute of System Dynamics and Control (SR) at the German Aerospace Center (DLR), Münchener Straße 20, 82234 Weßling, Germany
e-mail: antoine.pignede@dlr.de

S. K. Saha and K. Rama Krishna (eds.), *Recent Developments in Multibody Dynamics*, Lecture Notes in Mechanical Engineering,
https://doi.org/10.1007/978-981-97-7525-5_2

in-situ mission is the Mars Science Laboratory "Curiosity". But even for this powerful and maneuverable rover some areas remain not accessible. Unfortunately, these are often places of considerable scientific interest. For example, caves or lava tubes where traces of life could be detected.

Wheeled locomotion is efficient on relatively smooth and steady ground but quickly reaches limitations, e.g. on boulder fields or soft soil. The Mars Exploration Rover "Spirit" for example got stuck and had to be abandoned. Legged locomotion on the other hand, while it could access many sites, has drawbacks as well. It introduces higher complexity and therefore potential places of failure. A wheeled rover can move even with a defect drive, a defect drive in a legged robot usually is end of mission.

This is why DLR's Institute of SR researches an alternative locomotion concept that is capable to traverse difficult terrain, is robust and intelligent by design. The central idea is to use one variant of hybrid locomotion between rolling and walking: the *rimless wheel*. This is beneficial for exploration of narrow places because the robot is able to climb obstacles that are very high compared to the rover size. Steep slopes or very soft sand fields can also be crossed reliably. See [1] for the overall mission concept of the Scout rover, Fig. 1 for the rover in the multi-body simulation environment and Fig. 2 the first prototype. The remainder of this text will concentrate on the simulation-driven development for the locomotion subsystem of the rover.

Robots with rimless wheels already exist, see the state of the art, but the presented rover has flexibility in the wheel and modularity in the whole rover concept. The Scout rover is composed of *segments* that are compliantly connected together. A segment is a box with drives, electronics, space for payload, etc. around an axis. Compliance is introduced to enhance the locomotion capabilities and shock absorption as well as to reduce the energy requirements.

The development of an advanced robotic system that has to meet tight volume and mass requirements while being robust and powerful, asks to balance possibly conflicting parameters. This naturally leads to apply optimization at several stages of development. The evaluation step of a rover optimization process is the simulation of a digital model that tests how well the parameter set at hand meets the requirements.

Fig. 1 The full Scout rover simulation

Fig. 2 The Scout rover prototype [2]

Thus, a detailed and accurate model of the rover is needed. Such a model also has the benefit to further reduce the time and costs because many things can be tested and simulated prior to set a physical prototype to potential risks.

2 State of the Art

2.1 *The Rimless Wheel*

The rimless wheel is not a new idea and has been studied in great detail previously. McGeer [3] is one of the first articles to investigate it, as one possibility for passive dynamic walking. Interestingly the preliminary conclusion is "that rim removal is not a very progressive modification of an ordinary wheel" [3]! Nevertheless, the idea was not engraved and analysis continued, e.g. in [4] one of the first articles entirely dedicated to the rimless wheel. Up until then, the spokes were considered to be rigid bodies.

The mobile robots *RHex* [5] and *Whegs* [6] first used rimless wheels with compliant spokes. After these successful first research attempts, the concept was reused and improved several times, see [7] for a comprehensive list.

2.2 *Modeling, Simulation, and Optimization*

Modeling is the process of describing a physical *system* as a closed set of laws that shall reproduce the system's behavior [8]. *Simulation* applies algorithms to the model to do experiments that shall predict the real system's behavior and help the engineer to gain knowledge about it [9].

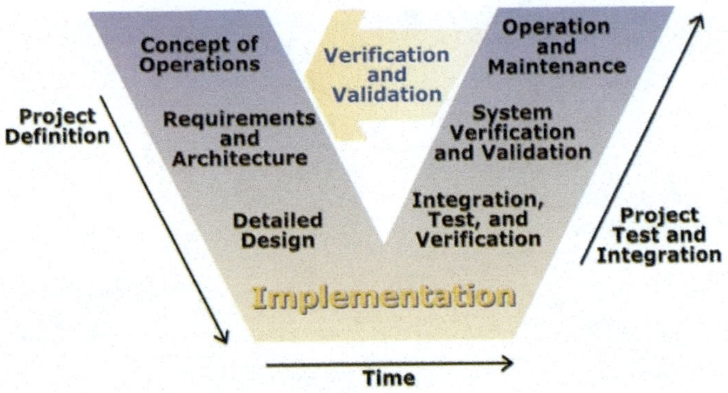

Fig. 3 The V-Model of development processes [12]

Modeling and simulation are often implemented as a V-shaped process, [10, 11] (German and US standards), Fig. 3. Beginning on the upper left side, the overall concept is tested with simple models. Further down, the complete system is broken down into components that are modeled with increasing level of detail. At the bottom of the "V", all subsystems are well understood and it is made sure that no important factor or disturbance is forgotten. This is also the place for verification and validation of the submodels. Going up the right side the subsystem models are integrated into the complete system model. This can be a relatively easy task if breakdown has been done well with clear interface boundaries. At the upper right side of the "V", the simulation model is complete and ideally represents the true physical system. In reality however the "V" is iterated several times or backward steps are taken until the verification and validation of the complete system is satisfactory.

Optimization in general consists of minimizing some cost function of design variables subject to constraints. There exist a large number of numerical optimization algorithms, see for example [13], and a possibly even larger amount of software packages, e.g. the widespread MATLAB Optimization Toolbox, Wolfram Mathematica or the free and open source SciPy.

Optimization can be done at almost any place of the "V" of Fig. 3. For example, to explore the capabilities and narrow down the parameter space early in the development. In the middle phases, the subsystems are optimized and in the later development trade-offs between subsystems can be resolved.

2.3 *Modeling, Simulation, Optimization for Planetary Exploration Rovers*

Modeling and simulation are not new to planetary rovers. A notable example is the *Adams-based Rover Terramechanics and Mobility Interaction Simulator ARTEMIS* developed for the Mars Exploration Rovers. Terramechanics play a central role to estimate the terrain traversing capabilities, but only rigid wheels are analyzed, whereas the Scout rover wheels have flexible parts. ARTEMIS has gone through a well-documented validation process and recently has also been extended to the discrete element method [14–16].

Optimization has been used to improve rover designs prior to this work too. However, there are two important differences in the process described here.

Usage of multibody dynamics simulation tools in Scout optimization [17, 18] are two examples of how modeling, simulation, and optimization have been used to develop and optimize rovers. Their rover simulation models used for the optimization are 2D quasi-static and not 3D fully dynamic simulation models, as those presented in this work. Michaud et al. [18] also developed a 3D fully dynamic simulation model, but it was not used in the optimization. The 3D model was only used after the optimization to verify the improved rover performance. References [17–19] inspired the optimization criteria and objective function Eq. 1. Ullrich et al. [20] motivated to use a genetic optimization algorithm.

Central position of modeling, simulation, optimization in the Scout development Modeling, simulation, and optimization of the Scout rover have many similarities with [21]. The same tools are used for modeling and optimization and a genetic algorithm is applied. Differences with [21] are exclusion of control parameters in the Scout tuner set. On the other hand, the detail in the multibody simulation is higher to capture the inherent advantages and drawbacks of the compliant rimless wheel rover concept.

This marks the second important difference of the Scout rover development with the process of other planetary exploration rovers: Modeling and simulation are major drivers. At DLR SR, this paradigm is often applied as *simulation driven development*. The models mentioned in Sect. 2.2 are not hardware prototypes but digital simulation models. A first prototype is only built once the first whole V-iteration is completed. Models are verified and usually a new "V" loop is triggered. As will become obvious in the next sections, this approach has proven to rapidly lead to good results.

3 Methods and Tools

Dymola and the Rover Simulation Toolkit The most important tool is *Dymola*, an environment for modeling and simulation of complex physical systems using the *Modelica* modeling language. The philosophy is to describe the behavior not in terms of algorithms but in terms of equations representing the true physical laws. Models are built hierarchically and in object-oriented way. The models are then simulated with one of the available numerical integration methods. Dymola is widely used at DLR SR. Scout is one rover building upon the in-house *Rover Simulation Toolkit RST* [22], a library of useful models for rovers that also extends the multibody mechanics with contact dynamics [23].

Discrete Element Simulation with Partsival While Dymola and RST is a good tool for multibody mechanics simulations, it has limitations for high-resolution contact dynamics. For example, when a Scout spoke is in soft sandy soil. The most precise way to do terramechanics is *Discrete Element Simulation* (or *Discrete Element*

Method, *DEM* for short) where a collection of rigid bodies representing the wheel are simulated in a bed of particles, each representing a pack of grains of the soil. Modern graphics cards have exceptional capabilities for DEM simulations and the SR tool *partsival* [24] makes good use of this and has advanced to one of the leading software packages in this field. It has been essential in the development of the rimless wheels for Scout.

Optimization Tools: Dymola and MOPS DLR SR has a large and powerful in-house optimization tool *MOPS* [25]. Its main environment is MATLAB but there also exists a library for Dymola [26]. Both are used in the rover development. In general, short and simple optimizations are done using the Dymola library while longer and complex campaigns use MATLAB (calling the Dymola simulation model in the evaluation step) because of the wider design choices and the advantages of powerful MATLAB computation and analysis tools.

4 Modeling the Scout Rover

4.1 Detailed Modeling of the Rimless Wheel in Partsival

The key component of the Scout rover is the rimless wheel with compliant spokes and *feet*. These are also equipped with "grousers" for higher stiction. Detailed investigation started the detailed definition phase. Many spoke designs were simulated, the foot shape was optimized and a study about the number of spokes was done. The main tool for this was partsival, the optimized rimless wheel has three spokes with a curved shape. The flexibility lowers the energy demand when turning counterclockwise in Fig. 4, which is beneficial for relatively flat ground. This sense is also often

Fig. 4 The rimless wheel analysis in the discrete element simulation *partsival*. 300,000 particles for 3 m are simulated, dark particles are inactivated by the dynamic boundaries [24]

best to cross obstacles. If the soil is very smooth the rover advances by turning the wheel clockwise and "swimming" through the sand as seen in Fig. 4. Following this computerized development, prototype wheels and feet were manufactured and their properties measured [7], the basis of the complete model.

4.2 Complete Rover Model in Dymola

Modeling and analysis of the rimless wheel in partsival belong to the "Detailed Design" phase of the V-Model of Fig. 3. The modeling of the complete rover in Dymola begins the "Implementation" phase but also partly the right branch of the "V" because results from the detailed designs are integrated into a common system.

There are possibilities to simulate flexible bodies in Dymola, however for simplification of the full system model and speedup of the simulation the spoke is represented as two rigid bodies with a linear torsion spring in between as depicted in Fig. 5. Likewise, the foot flexibility is modeled with a linear translational spring between the end of the spoke and the mounting of a rigid foot model. This property is shown in Fig. 5 but for clarity not in the other sketches. Simplifying the flexible bodies to linear springs is valid because the spoke and feet prototypes have a long range of linear stiffness and damping behavior as was experimentally identified [7].

In parallel with the spoke and foot springs are linear dampers. The damping constants d are calculated using the spring stiffness c and a damping ratio ζ (material parameter) after the well-known formula $d = 2\zeta\sqrt{mc}$ with the foot or spoke mass m respectively. All values are listed in Table 1.

As shown in [7] the best number of spokes is three, spaced by 120°. This is the best compromise between ability to climb high obstacles and traverse very soft terrain

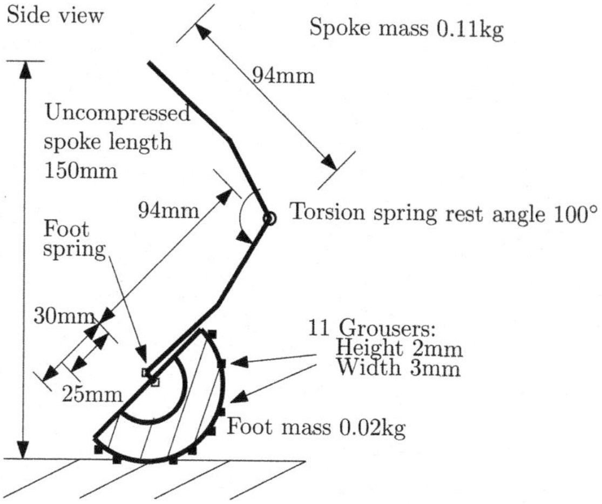

Fig. 5 Sketch of a spoke

Table 1 Most important Scout parameters

Parameter	Value
Total mass	18 kg, of which 6 kg for payload
Distance between segments (COM)	0.4 m
Spring stiffness between segments	30 N m rad^{-1} (roll), 10 N m rad^{-1} (pitch)
Damping ratio between segments	0.4 (roll and pitch)
Segment width, length and height	0.35 m × 0.2 m × 0.1 m
Middle segment length	0.22 m
Total wheel mass	0.6 kg
Hub diameter	0.06 m
Hub length	0.044 m
Uncompressed spoke length	0.15 m
Spoke torsion spring constant	35 N m rad^{-1}
Spoke damping ratio	0.1
Spoke material friction parameter	0.45
Foot radius	0.03 m
Foot width	0.057 13 m
Foot spring constant	10 000 N m^{-1}
Foot damping ratio	0.2
Foot material Young's modulus	1.24×10^5 Pa
Foot material friction parameter	0.7
Grousers	0.002 m (height), 0.003 m (width)
Angle between grousers	0.2618 rad = 15°, there are 11 grousers

while reducing the undesired motions induced by removing the rim. Figure 6 summarizes this in two sketches from the front and side.

The assembly of two wheels and the box in between is called *segment*, sketched in Fig. 7. Segments are connected by flexible plates, hereafter called *vertebrae* and depicted in Fig. 8. Here again, a linear torsion spring between two rigid bodies is used as a model in Dymola for simplification reasons. It has two rotational degrees of freedom: along the rover traveling direction (roll) and parallel to the axis of the wheels (pitch). For the rover prototype, plates made out of polyoxymethylene (POM) are used which allow for flexibilities in the range of about 10–100 N m rad^1 depending on the plate thickness, shape and manufacturing method. Determining an optimal value will be subject of the optimization campaign, see Sect. 5.1.

The damping of a vertebra is calculated using a damping ratio as the spoke and foot damping constants. The POM material has $\zeta = 0.4$.

Important parameters are summarized in Table 1. Some parameters are still subject to change in optimizations (e.g. the springs between segments, the width of the segments) or due to constraints (e.g. the length or width because of the motor or payload choice). The wheel parameters and the total mass limit of 18 kg will remain.

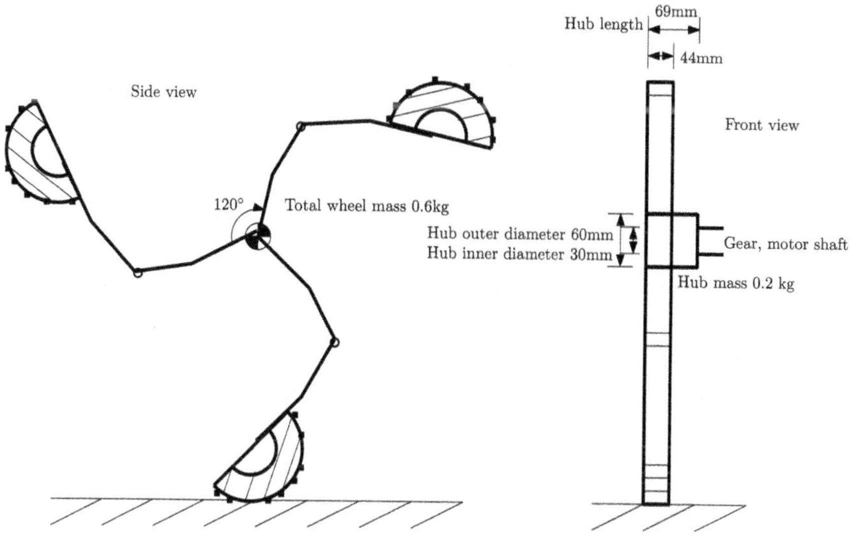

Fig. 6 Sketch of a wheel

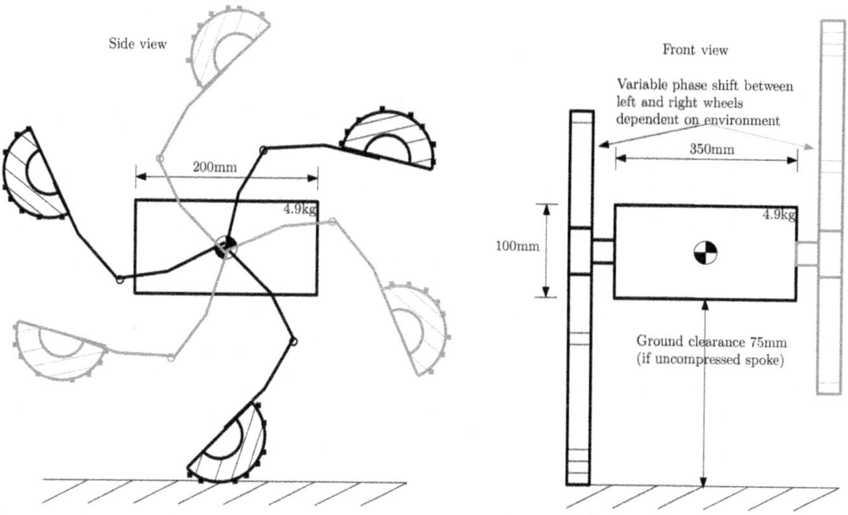

Fig. 7 Sketch of a segment

As visible in Fig. 7, the phase shift between wheels on the same segment is variable, likewise for the phase shifts between wheels on the same rover side but different segments. The already mentioned biologically motivated robot RHex has shown that different kinds of phase shift sets can be beneficial depending on the terrain geometry [27]. Intuitively, on flat terrain with low obstacles, a maximum phase

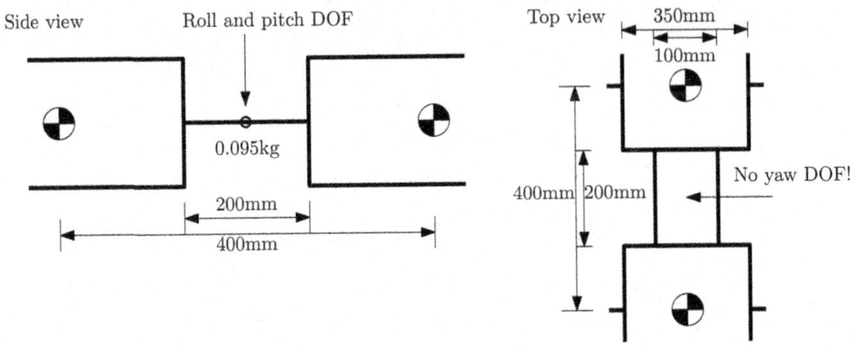

Fig. 8 Sketch of a vertebra

difference is good (energy savings and less shaky drive). For crossing large and high obstacles, it is better if the wheels are in phase to make use of the curved spoke shape. This counts however only for forward driving. During skid steering and point turns, the phase shift cannot be kept constant because wheels on one side move faster than on the other side. The selection of suitable phase shifts as well as the design to control the shifts is not a trivial task. It is the subject of a dedicated publication [28]. The Scout simulation model proved to be of good use in this process.

5 Applications

5.1 *Optimize the Vertebra Stiffness*

To enhance the locomotion performance of the Scout rover, the segments are connected using flexible vertebrae. They allow movement in two degrees of freedom. *Roll* is along the rover driving axis and *pitch* is parallel to the wheel axes. The length of the vertebra plates is determined by the spoke length (spokes of adjacent segments mustn't touch each other) and the overall rover length. In contrast, the flexibilities along roll and pitch are free within the relatively large manufacturing constraints. This is a good example where one can perform an optimization campaign with the simulation model. This requires the definition of the parameters to optimize, the criteria to measure the performance and the optimization algorithm to use.

Tuners Parameters to optimize are often called *tuners*. Here, these are the stiffness values of torsion springs along the roll and pitch axes of the vertebrae. The bounds are 10–100 N m rad[1] for both degrees of freedom. The damping is not a tuner because the damping ratio 0.4 is constant for the chosen material and stiffness range.

Scenario and criteria The *criteria* and *objective* quantify how well the tested tuner values perform. These words are sometimes used interchangeably, but they're not exactly equal. *Criteria* or *metrics* are the individual evaluated values, while *objective* or *fitness* refer to the collection of criteria, e.g. absolute maximum criterion or sum of squares of all criteria.

The objective is that the Scout rover successfully crosses several obstacles, preferably with minimal energy and torque demand on the drives. The obstacles to cross set a *scenario*, that is a simulation run in which the criteria are calculated to measure the fitness of the current choice of vertebra stiffness. The obstacles that the Scout rover must overcome are stairs, slopes, steps, and ditches. They're depicted in the screenshots of the Figs. 9, 10, 11, and 12. A fifth scenario part is a flat plane

Fig. 9 Stairs obstacle

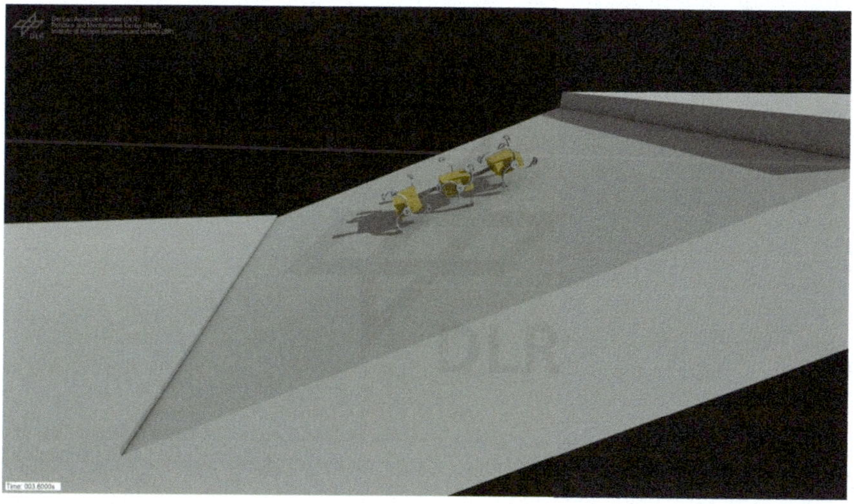

Fig. 10 Slope obstacle

Fig. 11 Broad steps and ditches

Fig. 12 Narrow steps and ditches

where the rover is commanded to turn the wheels with 10 rad s^{-1} instead of 4 rad s^{-1}. No curves are driven and the phase difference between wheels on the same axis is 60°.

The "swimming" case (see also Fig. 4) is not included in the optimization because the simulation of the soft sand demands too much computation power. In addition, there is only little pitch and roll in the vertebrae in "swimming" when the segment boxes lie on the ground or very close to it.

Throughout the scenario, the total energy needed to finish the course is summed and the torque on the wheels are recorded. The objective function is

$$f(c_r, c_p) = \left(\frac{\sum E(c_r, c_p)}{\sum E(15, 60)}\right)^2 + \left(\frac{\max \tau(c_r, c_p)}{\max \tau(15, 60)}\right)^2 + \left(\frac{\bar{\tau}(c_r, c_p)}{\bar{\tau}(15, 60)}\right)^2 \quad (1)$$

(sum of squares) where c_r and c_p are the tuners, $\sum E$ is the total electrical energy, $\max \tau$ is the highest torque peak recorded anywhere and $\bar{\tau}$ is the arithmetic mean over all motor torques over the whole simulation time. These are normalized with the criteria from the initial tuners (arguments (15, 60)). Only evaluations where all obstacles where successfully crossed are retained.

Optimization algorithm MOPS is used to do the optimization, a genetic algorithm with a population of 20 individuals and a maximum of 200 generations. Elitism (survival of the fittest only), no niching, 0.8 crossover probability and 0.05 mutation probability are selected.

Results The start values of the vertebra spring stiffnesses are $15\,\mathrm{N\,m\,rad^{-1}}$ in roll and $60\,\mathrm{N\,m\,rad^{-1}}$ in pitch, these were determined manually from the forward driving performance on a flat plane. There a stiff pitch spring and loose roll spring are beneficial to profit from the compliant spokes that store energy when wheels on the same axis have maximum a phase difference.

However, it turns out that the optimal spring stiffnesses for the Scout rover to cross the obstacles are $30\,\mathrm{N\,m\,rad^{-1}}$ in roll and $15\,\mathrm{N\,m\,rad^{-1}}$ in pitch. A stiffer roll spring prevents the rover from diving too much for the case depicted in Fig. 12. Likewise, a lower pitch spring stiffness is better for climbing stairs as depicted in Fig. 9, especially for the beginning and ending.

Figure 13 shows the progress of the optimization over the 200 generations. The tuners, criteria, and objective of the currently fittest individual are plotted. The tuner progress (left plots of Fig. 13) shows a typical pattern for a genetic algorithm where at first the values vary a lot before reaching an optimal individual about at the middle of the generation numbers. Thereafter the genetic algorithm doesn't find any further

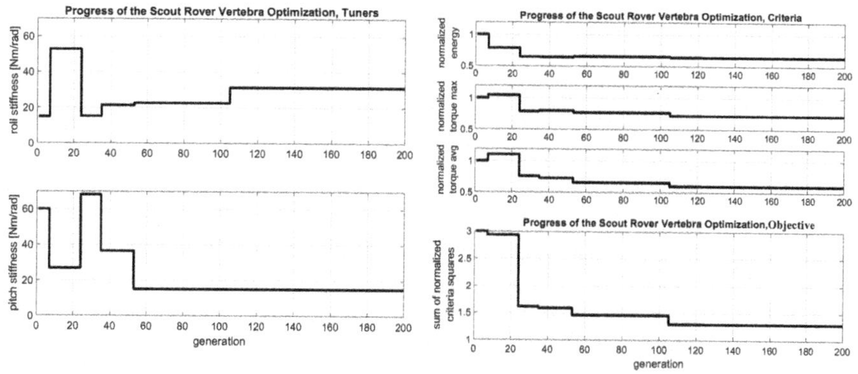

Fig. 13 Optimize the vertebra stiffness: tuners, criteria, and objective

improvements. The normalized criteria progress (right plots of Fig. 13) shows that improvements in all three criteria are found. The first generations find tuners that are better in terms of energy at the cost of higher torques, but later in the optimization process roll and pitch stiffness pairs are found that also reduce the torque demands on the drives. The larger plot on the lower right side of Fig. 13 is the application of the sum of squares objective function on the criteria (Eq. 1).

5.2 Determine the Maximum Drop Height

The Scout rover is built robustly to survive high drops that it most likely is going to encounter while exploring caves [1]. Finding out the drop height that the rover can survive is much easier, cheaper, and safer with the simulation.

Again, the soft soil case is not investigated. Only shocks on flat and hard ground with relatively high Young's modulus are done because drops on reasonably flat and hard underground is what to expect in tubes of solidified lava. In addition, soft soil dissipates more energy on impact. In other words: If the rover survives a drop from some height on this kind of ground, it will certainly survive it as well on soft soil. Uneven terrain or the case where a segment box hits a sharp rock are deliberately neglected.

The commands to the drives are another important part of the scenario definition. One possibility to make shocks less harmful is to let the drives loose. However, in the realistic case that the rover drives on its own and then falls over an edge, the drives are going to continue turning. This is why three different cases are examined.

- The rover falls from a variable height on a flat plane, the drives are locked. The spokes point straight down in order to have the maximum compression length before the segment boxes touch the ground and thus the best protection of the payload.
- The rover falls from a variable height on a flat plane, the drives are free, i.e. no active torque opposes the movement of the wheels.
- The rover drives forward and falls down to the flat plane underneath, the wheels turn with the same speed throughout the simulation.

Ten simulations with height from 1 to 10 m in Earth gravity are done.

The fourth-order Runge-Kutta method is used with the fixed step size 0.1 ms for heights up to and including 5 m and 0.01 ms for heights above 5 m. Tests with variable step solvers simulate in about the same time and show no significant differences. Note also that material fracture or plastic deformation is not modeled.

Results The aim of the campaign is to find out how high a drop the Scout rover can survive. The plots in Figs. 14 and 15 summarize the results. They show the highest recorded radial and axial force on the wheel shafts, the parts most likely to break first on impact. Further plots show the highest recorded force on the segment boxes.

For the two first categories, when the rover lands on the six feet at the same time, forces increase linearly with height and are higher on the middle segment than on the

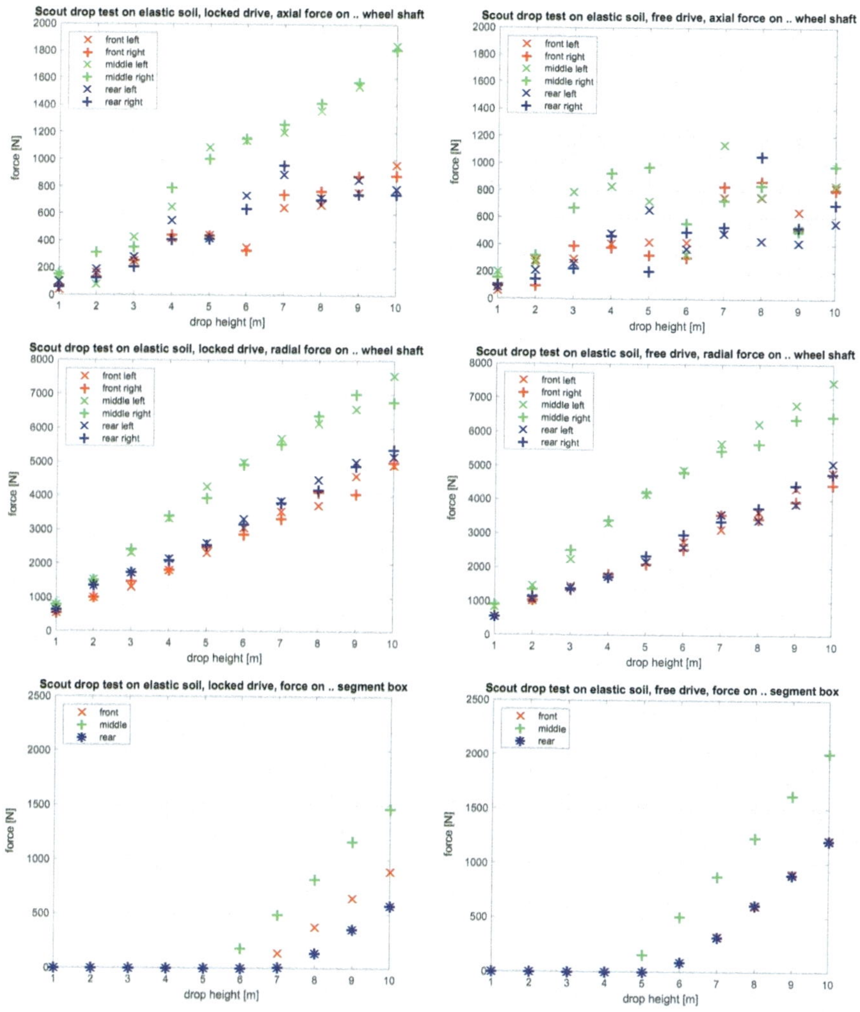

Fig. 14 Drop test results: force peaks on wheels (axial on the top, radial in the middle) and on segment boxes (bottom), with locked (left) or free (right) drives

front and rear segments (always almost equal values), because of the pitch degree of freedom in the vertebrae. Comparing the results of the locked and free drive (left and right in Fig. 14), forces on the wheel shaft are lower and forces on the segment boxes are higher. This is expected as the motivation to have free drives was to reduce the loads on the shaft, possibly at the cost of a higher impact on the segment box.

The middle segment box doesn't touch the ground for drops heights up to 5 m. The front and rear boxes don't touch the ground for drops heights up to 6 m. Critical forces are only attained at higher heights. 5–6 m is also the height where radial loads on the drive become fatal. The critical force value (and thus the critical drop height) is

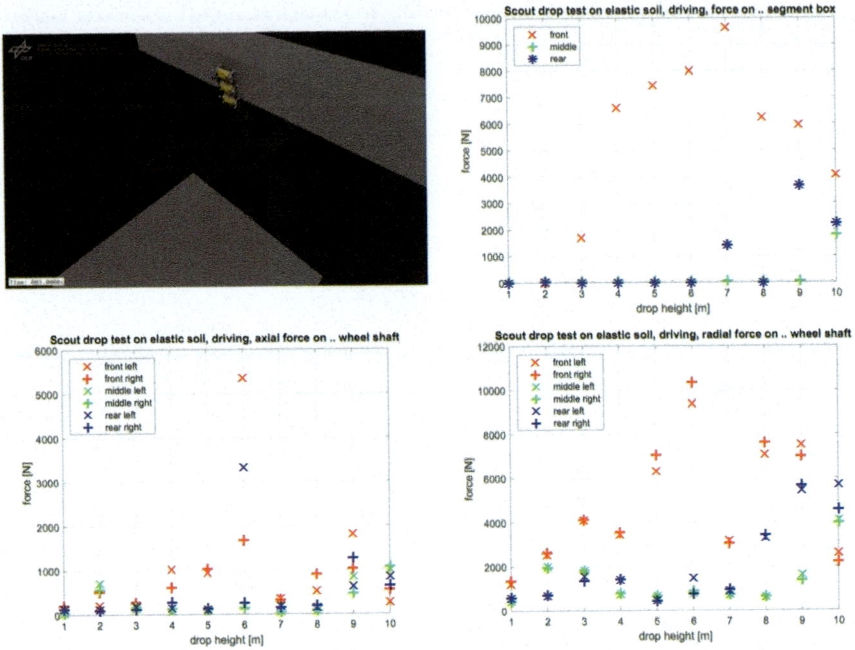

Fig. 15 Driving drop test: simulation screenshot (10 m height) and force peaks on segment boxes and on wheels, axial and radial components

difficult to capture in a single figure. The motor and gear data sheets have maximum axial and radial force, a little lower than 1000 N, but due to the way in which the rover is built and the shortness of the shocks, higher forces are probably survivable.

The height at which drops become dangerous, is almost not affected by the mitigation attempt to leave the drives free. This is a good result because there is just little time for the rover to detect that it falls and react accordingly.

The plots of the more realistic case when the Scout rover drives and falls down some height on a flat plane, Fig. 15, are more difficult to analyze.

When the rover drives over the edge, a pitch rotation is induced. At heights of 6 m or more, the rover ends upside down after the short bouncing phase. Because of this rotary movement, the front segment experiences the highest loads and hits the ground already when the drop height is 3 m. Moreover, the force values are much higher, compare the bottom plots of Fig. 14 with the middle right plot of Fig. 15. The wheel shafts experience comparable loads than in the first two simulation categories.

The conclusion of this simulation campaign is that the rover can survive drops of *up to* 3 m *height* in Earth gravity on relatively flat terrain. If the pitch angle is low, 5 m drop height is still safe. The critical parts of the rover are the wheel shafts which have to take high loads and are more fragile than the spokes and feet. However, in a more realistic scenario where the pitch angle is high, such that the front or rear segment is exposed, more than 3 m is fatal because the box will probably be damaged.

6 Conclusion and Further Work

This article shows how modeling and simulation embedded in the V-shaped process (Fig. 3) are successfully implemented to develop a rover with rimless wheel that is robust and uses flexibilities and an unconventional wheel advantageously. This makes the Scout rover capable to reach area of scientific interest on the surface of Mars or the Moon such as caves or lava tubes. A detailed model of the modular rover is presented (Sect. 4) and the performance is significantly improved by simulation and optimization (Sect. 5.1). More applications of modeling, simulation, and optimization justify the term *simulation driven development*, an example is given in Sect. 5.2.

Table 2 lists some requirements for the rover and compares them with values achieved by the simulation model. The last two rows are not requirements, but unexpected abilities of the prototype, reproducible in simulation. When the rover lies on a side, turning the wheels induces enough instability to make it tumble and be on its feet again. The rover remains maneuverable with five operating wheels should one drive experience malfunction. Similarly, the performance is not significantly reduced when up to three spokes are broken.

The approach of this text is a comparatively fast and cheap way to design and analyze a rover. 1.5 years after the start of the project, when the simulation model had gone through an iteration of the V-model, the first hardware prototype was assembled. Thanks to the optimizations and the experience gained from the modeling and simulation, it already exceeds the requirements and expectations, often even outperforming the simulation. For example, the prototype is able to climb slopes up to 30°.

The prototype and simulation models generate different results for the moment. Verification and validation are important, when modeling, simulation, and optimization are major drivers of the development process. So far, only some aspects such as the spoke and feet flexibilities have been modeled after hardware. To continue to reduce the differences in the capabilities of the prototype and simulation model, hardware tests and measurements of subsystems, e.g. the drives, are going to be done. This likely will trigger a new iteration of the V-Model for the Scout rover simulation.

Table 2 Requirements for the Scout and performance of the simulation

Parameter	Requirement	Simulation
Maximum obstacle height	>0.15 m	0.25 m
Maximum slope (sand)	>20.0°	20.0°
Maximum slope (rocky terrain)	>20.0°	20.0°
Maximum drop height (Earth gravity)	>1.0 m	3.0 m
Upside down driving	✓	✓
Traversing soft ground	✓	✓
Uprighting from side	N/a	✓
Driving with one defect wheel	N/a	✓

The Scout rover is expected to be ready for a planetary exploration mission around 2030 [1]. Until then, modeling, simulation, and optimization will continue to accompany the development of the rover. These activities will also be valuable during the operation phase and for analysis of the downlink data.

References

1. Lichtenheldt R et al (2021) A mission concept for lava tube exploration on Mars and Moon—the DLR Scout rover. In: 52nd Lunar and planetary science conference, 15–19 March 2021. LPI
2. Wiesner M (2018) Design and control of a planetary exploration Scout for extreme environments based on the rimless wheel. Master's thesis, Universität der Bundeswehr München
3. McGeer T (1990) Passive dynamic walking. Int J Robot Res 9(2):62–82
4. Coleman MJ et al (1997) Motions of a rimless spoked wheel: a simple three-dimensional system with impacts. Dyn Stabil Syst 12(3):139–159
5. Altendorfer R et al (2001) RHex: a biologically inspired hexapod runner. Auton Robots 11(3):207–213
6. Quinn RD et al (2002) Improved mobility through abstracted biological principles. In: International conference on intelligent robots and systems, 30 September–5 October 2002, Lausanne. IEEE/RSJ, pp 2652–2657
7. Stubbig L (2017) Model-based development of a mobile scout for extreme terrain in planetary exploration. Bachelor's thesis, Technische Universität Ilmenau
8. Cellier FE (1991) Continuous system modeling. Springer, New York
9. Cellier FE, Kofman E (2006) Continuous system simulation. Springer, New York
10. Bundesrepublik Deutschland (2006) V-Modell-XT: Das deutsche Referenzmodell für Systementwicklungsprojekte, 2.3 edn. Munich
11. National ITS Architecture Team (2007) System engineering for intelligent transportation systems: an introduction for transportation professionals. Washington
12. After Osborne L et al (2005) Redrawn by User: Slashme. Image extracted from Clarus concept of operations. Publication No. FHWA-JPO-05-072, Federal Highway Administration. https://commons.wikimedia.org/wiki/File:Systems_Engineering_Process_II.svg. Public domain
13. Nocedal J, Wright SJ (2006) Numerical optimization, 2nd edn. Springer series in operations research and financial engineering. Springer, New York
14. Iagnemma KD et al (2011) Terramechanics modeling of Mars surface exploration rovers for simulation and parameter estimation. In: 8th International conference on multibody systems, nonlinear dynamics and control, 28–31 August 2011, Washington. ASME, pp 805–812
15. Zhou F et al (2014) Simulations of Mars rover traverses. J Field Robot 31(1):141–160
16. Johnson JB et al (2015) Discrete element method simulations of Mars exploration rover wheel performance. J Terramech 62:31–40
17. Krebs A et al (2006) Performance optimization of all-terrain robots: a 2D quasi-static tool. In: International conference on intelligent robots and systems, 9–13 October 2006, Beijing. IEEE/RSJ, pp 4266–4271
18. Michaud S et al (2006) Rover Chassis evaluation and design optimisation using the RCET. In: 9th Workshop on advanced space technologies for robotics and automation, 28–30 November 2006, Noordwijk. ESA
19. Lachat D et al (2006) Antarctica rover design and optimization for limited power consumption. IFAC Proc 39(16):788–793
20. Ullrich F et al (2010) Design optimization of a Mars rover's rocker-bogie mechanism using genetic algorithms. In: 10th Australian space science conference, 27–30 September 2010, Brisbane. SASIC, pp 199–210

21. Seeni A et al (2010) Conceptual rover design by an integrated optimization process. In: 10th International symposium on artificial intelligence, robotics and automation in space, 29 August–1 September 2010, Sapporo. ESA, pp 538–545
22. Hellerer M et al (2017) The DLR rover simulation toolkit. In: 14th Symposium on advanced space technologies in robotics and automation, 20–22 June 2017, Leiden. ESA
23. Lichtenheldt R et al (2016) Wheel-ground modeling in planetary exploration: from unified simulation frameworks towards heterogeneous, multi-tier wheel ground contact simulations. In: Font-Llagunes JM (ed) Multibody dynamics. Springer, pp 165–192
24. Lichtenheldt R et al (2018) Partsival—collision-based particle and many-body simulations on GPUs for planetary exploration systems. In: 5th Joint international conference on multibody system dynamics, 24–28 June 2018, Lisbon. Springer
25. Joos H-D (2002) A multi-objective optimisation-based software environment for control systems design. In: Conference on computer-aided control systems design, 18–20 September 2002, Glasgow. IEEE, pp 7–14
26. Pfeiffer A (2012) Optimization library for interactive multi-criteria optimization tasks. In: 9th international Modelica conference, 3–5 September 2012, Munich. Linköping University Electronic Press, pp 669–680
27. Campbell D, Buehler M (2003) Stair descent in the simple hexapod 'RHex'. In: International conference on robotics and automation, 14–19 September 2003, Taipei. IEEE, pp 1380–1385
28. Skibbe J (2021) Analysis of phase shifts for a rimless wheel rover. In: Aerospace conference, 6–13 March 2021. IEEE

Simulation and Control of Shape Memory Alloy Spring Actuator in a Flexible Tube Manipulator

Nisha Bhatt, Vedanshu Seedwan, Samyak Jain, Sanjeev Soni, and Ashish Singla

Abstract The current article describes the modeling and positional control of a flexible tube manipulator intended for medical applications. The flexible tube is made of polyamide material and its tip is controlled via a calibration-based controller. The configuration is made compact with Shape Memory Alloy (SMA) spring pair as the prime actuators. Two antagonistically connected SMA spring pairs manipulate the flexible tube in 2D and the amount of current required to control the spring is delivered via the controller. The dynamics of the SMA spring pair along with the flexible tube dynamics are presented and the simulation predicts the force transmission capability of the actuator. Thereafter, experimental analysis shows the positional tracking ability of a flexible tube manipulator.

Keywords SMA actuator · Flexible tube manipulator · Calibration controller · Positional tracking

1 Introduction

Robotics is still in its evolution phase and in recent years robotics technology has been widely applied in production industries. A typical industrial robot consists of rigid links that handle objects around the assembly chains. Unlike these rigid counterparts, there is a new concept of robotic manipulators known as continuum robots which have proven their potential in many research applications. While the concept is still being studied, their popularity is growing every day due to their lower weight, increased flexibility, dexterity, and more degree of freedom as compared with their rigid counterparts. Generally, continuum manipulators consist of a flexible structure made up of a single segment or multiple segments. These flexible structures

N. Bhatt (✉) · V. Seedwan · S. Jain · A. Singla
Thapar Institute of Engineering and Technology, Patiala 147004, India
e-mail: nbhatt20_phd17@thapar.edu

S. Soni
Central Scientific Instruments Organization (CSIR-CSIO), Chandigarh 160030, India

S. K. Saha and K. Rama Krishna (eds.), *Recent Developments in Multibody Dynamics*, Lecture Notes in Mechanical Engineering,
https://doi.org/10.1007/978-981-97-7525-5_3

can be actuated via pneumatics [1], hydraulics [2], cables [3], tendons [4, 5], electric motors [6], and smart material actuators [7, 8]. Continuum manipulators have been developed through a new era with the advent of smart materials.

Smart materials have recently been popular due to their high potential to fortify engineering applications. Among smart materials, shape memory alloys are quite popular owing to their large recoverable strains due to crystal structure's transformations between martensite and austenite phases. This change in crystal structure results due to a change in stress and temperature. At lower temperatures, they are deformable to a large extent. This deformation can be restored if heated above their transformation temperatures. During the transformation, many properties of the material change including electrical resistance. If these materials are used as sensors, a change in electrical resistance during the transformation can become a measure of the sensor. Similarly, if used as actuators, resistance to complete shape recovery will exert a force. These materials are widely used as actuators or sensors in different domains of applications that include aerospace [9], medical [10], construction [11], robotics [12], etc. Their applications in continuum manipulator development are the area of interest in this article.

In this paper, a silicon rubber tube is used as a continuum manipulator, which is actuated via two pairs of antagonistic SMA springs. The paper describes a mathematical model of the flexible tube which is simulated in MATLAB software and depicts the force-transmitting capability of an antagonistic spring pair. The simulation predicts the temporal variation of the bending angle of the tube in one plane when subjected to a given voltage pattern. To bend the tube in another plane, the same mathematics is used for another antagonistic SMA spring pair. Once the expected output generation capability of the manipulator is determined through simulation, the tip of the tube is experimentally made to follow a series of points via a calibration-based control algorithm. The experimental results of trajectory tracing are discussed for the tube manipulator.

2 Simulation and Control

The flexible tube manipulator comprises a silicon rubber flexible tube actuated via two pairs of antagonistic SMA springs. One end of all springs is fixed on the mounting and the other end is fixed by connectors on the metallic spacers placed near the tip of the flexible tube. A single SMA spring can provide actuation in one direction and requires a resetting mechanism to reach its initial position. To achieve bidirectional motion in an axis, another SMA spring is placed parallel to the first SMA spring which brings the SMA spring to its initial position in a fast manner during the cooling phase and provides the motion in the opposite direction to the first SMA. Thus, a single antagonistic pair provides motion in say X direction. To achieve motion in the Y direction, a similar antagonistic SMA spring pair is used. Thus, two pairs provide 2D motion to the flexible tube manipulator. Since the mathematics of one pair of antagonistic SMA spring actuation is similar to the other pair. Thus, only one pair is discussed in this article. In the pair of antagonist springs, the first SMA spring is heated by electrical heating (via multi-channel SMA amplifier module by *Spranktronics*, each channel rating 10 V, 2 A), which contracts an extended length of this SMA spring. At this instant, the other SMA spring acts as a bias to restore the contracted length of the first heated SMA to its initial position. Similarly, when the second SMA spring is heated, the first spring which has now been cooled acts as a bias to the second spring. During this actuation, the flexible tube which is in contact with the antagonistic SMA spring pair bends to an angle that is a function of the electric current passing through the spring pair at that instant. A preliminary setup of the proposed manipulator is shown in Fig. 1.

Initially, the force transmission capability of the actuator pair and bending angle variation are determined by simulating the mathematical model of the actuator on MATLAB. To develop the mathematics of the actuator, an existing constitutive model of the SMA spring based on loading history [13] is considered. The phase kinetics of the said model are adopted to develop the ordinary differential equations of the spring pair and simulated via an *Implicit solver* on MATLAB. The heat transfer model is adopted from the Banerjee model [14]. The following equations are used to develop a complete mathematical model.

$$C_{sp_1}\frac{\partial T_{sp_1}}{\partial t} - \lambda_{sp}\frac{\partial \varsigma_1}{\partial t} = \frac{V_1^2}{\rho_{sp_1} l_{sp_1}^2} - \frac{4h}{d_{sp}}\left(T_{sp_1} - T_a\right) \tag{1}$$

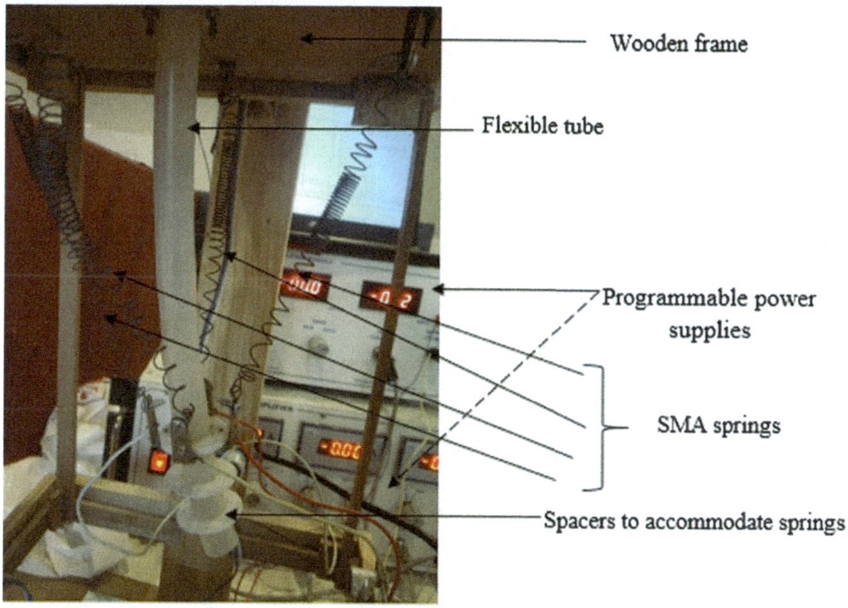

Fig. 1 Preliminary setup of flexible tube manipulator

$$\frac{\partial \tau_1}{\partial t} = -\left(G_1 \frac{d_{sp}}{\pi D_{sp}{}^2 N_{sp}} r_d \frac{d\theta_x}{dt} + G_1 \gamma_l \left(\frac{\partial \varsigma_{s1}}{\partial t}\right)\right) \tag{2}$$

$$\frac{\partial \varsigma_1}{\partial t} = F_{11} \frac{\partial T_{sp1}}{\partial t} + F_{21} \frac{\partial \tau_1}{\partial t} \tag{3}$$

$$\frac{\partial \varsigma_{s1}}{\partial t} = F_{31} \frac{\partial T_{sp1}}{\partial t} + F_{41} \frac{\partial \tau_1}{\partial t} \tag{4}$$

$$C_{sp2} \frac{\partial T_{sp2}}{\partial t} - \lambda_{sp} \frac{\partial \varsigma_2}{\partial t} = \frac{V_2{}^2}{\rho_{sp2} l_{sp2}{}^2} - \frac{4h}{d_{sp}} (T_{sp2} - T_a) \tag{5}$$

$$\frac{\partial \tau_2}{\partial t} = G_2 \frac{d_{sp}}{\pi D_{sp}{}^2 N_{sp}} r_d \frac{d\theta_x}{dt} - G_2 \gamma_l \left(\frac{\partial \varsigma_{s2}}{\partial t}\right) \tag{6}$$

$$\frac{\partial \varsigma_2}{\partial t} = F_{12} \frac{\partial T_{sp2}}{\partial t} + F_{22} \frac{\partial \tau_2}{\partial t} \tag{7}$$

$$\frac{\partial \varsigma_{s2}}{\partial t} = F_{32} \frac{\partial T_{sp2}}{\partial t} + F_{42} \frac{\partial \tau_2}{\partial t} \tag{8}$$

$$\frac{d^2\theta_x}{dt} + \frac{c_t}{I_d} \frac{d\theta_x}{dt} + \frac{E_t I_t}{l_t I_d} \theta_x = \frac{F_{net} r_d}{I_d} \tag{9}$$

where, C_{sp}, λ_{sp}, ρ_{sp}, l_{sp}, d_{sp}, D_{sp}, G, γ_l, T_{sp}, τ, ς, ς_s, V, h, N_{sp} and θ_x are specific heat capacity, latent heat capacity, resistivity, current length of spring, diameter of spring wire, mean diameter of spring, modulus of rigidity, maximum limit of shear strain in spring, temperature of spring, shear stress, martensite fraction, stress-induced martensite fraction, applied voltage, convective heat transfer, number of coils and bending angle of the tube in the x direction, respectively. The subscripts 1 and 2 correspond to SMA spring 1 and SMA spring 2, respectively. In addition, F_{1i} to F_{4i} are the phase kinetics function values that are mentioned in the literature [15]. Here, $i = 1, 2$ corresponds to springs 1 and 2, respectively. Table 1 shows the numeric values of the constants considered for simulation.

Once the force exerted by the antagonistic spring pair and corresponding angular movement of the flexible tube is estimated through the simulation, calibration of the springs and data acquisition are carried out. An accelerometer cum gyroscope (MPU 6050) sensor is attached to the tip of a flexible tube that determines the X and Y coordinates of the tube tip. Although the Z coordinate is also known through the sensor, its variation in tracking a two-dimensional shape is negligible. Thus, only the X and Y coordinates of the tube tip are considered. To develop the control algorithm, two calibrations are carried out: (1) calibration of sensor coordinates versus applied voltage and (2) calibration of voltage provided by the data acquisition card (USB-6001(2 AO), *National Instruments and my DAQ* (2 AO), *National Instruments*) versus voltage on the power supply (Multi-channel SMA Amplifier Module, *Spranktronics*). As both X and Y coordinates of the tube tip are required to trace a 2D shape, calibration

Table 1 Properties of flexible tube manipulator

Parameters	Value
Diameter of SMA spring wire (d_{sp})	0.75 mm
Mean coil diameter of SMA spring (D_{sp})	7.5 mm
Number of coils in SMA spring (N_{sp})	50
Ambient temperature (T_a)	26 °C
Resistivity of SMA (ρ_{sp})	0.009 Ohm-mm
Specific heat capacity (C_{sp})	0.0052 J/K-mm^3
Convective heat transfer coefficient (h)	180 W/m^2°C
Modulus of rigidity in martensite (G_m)	22.45 GPa
Modulus of rigidity in martensite (G_a)	30.03 GPa
Length of tube (l_t)	100 mm
Moment of inertia of spacer disc (I_d)	400 kg mm^2
Radius of spacer disc (r_d)	12 mm
Young's modulus of flexible tube (E_t)	0.5 GPa

of the sensor is carried out for both X and Y directions. Thus, two columns are made which contain values of X coordinate versus voltage required, and similarly, another two columns contain values of Y coordinate versus voltage required. All these values are saved in the software itself. Another calibration is carried out by shorting each channel of the power supply and sending the voltage value to the analog output pins of the above DAQ cards that correspond to the shorted supply channel. Thus, for the four analog outputs of the power supply, four columns are created that contain the voltage given to the DAQ card. Once sufficient calibrated data are obtained by carrying out the calibration task manually, a calibration module is created in LABVIEW software, which contains the above calibration data. In the main program, a series of setpoints (x and y coordinates) are provided to the controller that automatically estimates the required value of the applied voltage to be given to the power supply and voltage given to data acquisition analog output pins through the calibration module. The functioning of the controller is similar to an artificial neural network controller that requires some experimental data set to get trained and once trained, it uses the training data to estimate required voltage values for given setpoints. A flow diagram of the hardware setup and its working is shown in Fig. 2.

3 Results and Discussions

The results shown in Fig. 3 have been obtained using the set of Eqs. 1–9 in MATLAB and using the material properties as tabulated in Table 1. The solution of these differential equations represents a state vector that denotes values of temperature, shear stress, total martensite fraction, stress-induced martensite fraction, and the bending angle of the flexible tube. A triangular wave input voltage profile is applied

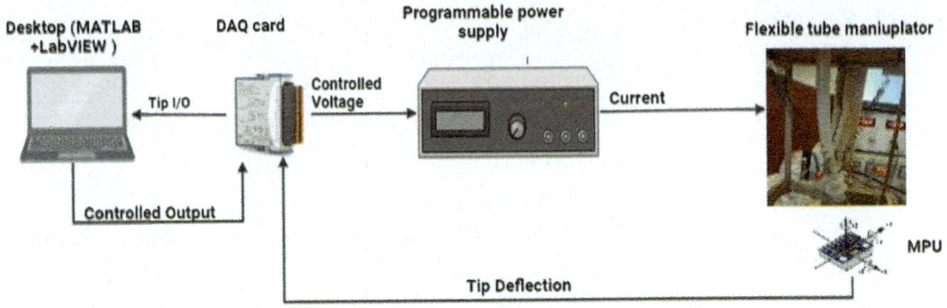

Fig. 2 Schematic of the flow of information in the complete setup of flexible tube manipulator

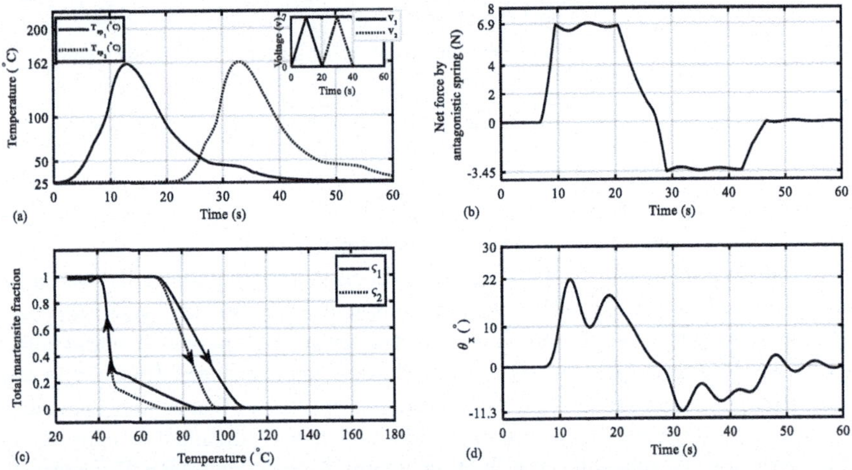

Fig. 3 Output response of antagonistic spring actuator: **a** temporal variation of temperature when subjected to a given voltage (inset), **b** temporal variation of the net force exerted by spring pair, **c** variation of total martensite fraction with temperature and **d** temporal variation of angular displacement of the flexible tube

to the first SMA spring for 20 s and then a similar voltage is given to the second SMA spring for the next 20 s, as shown in the inset of Fig. 3a. Due to this input voltage pattern, SMA spring 1 is subjected to a heating and cooling cycle and then SMA spring 2 is subjected to a heating and cooling cycle. The variation of temperature with time for both springs is shown in Fig. 3a. It is important to note that both springs are subjected to space constraints as their lengths are fixed by the spacers. Thus, free recovery of SMA spring is not possible. Under the constrained recovery, each SMA spring exerts some force that is opposed by its antagonistic counterpart. Thus, a maximum net force of 6.9 N is obtained in the positive direction and similarly, 3.45 N is obtained in the negative direction as depicted in Fig. 3b. The positive and negative signs indicate the directions in which the antagonistic springs generate force. The two signs indicate the bidirectional motion of the tube.

Table 2 Desirable setpoints and traced data by the manipulator

Vertices of square (setpoint)	Traced points at that instant (process variable)	Error = setpoint-process variable
(0, 0)	(0.29, 0.066)	(−0.29, −0.066)
(10, 0)	(9.93, 0.128)	(0.07, −0.128)
(10, 10)	(9.90, 9.67)	(0.1, 0.33)
(−10, 10)	(−5.10, 9.97)	(−4.9, 0.03)
(−10, −10)	(−9.96, −8.63)	(−0.04, −1.37)
(10, −10)	(9.91, −9.93)	(0.09, −0.07)
(10, 0)	(10.08, 0.11)	(−0.08, −0.11)

Since the obtained temperature and force values lie within the transformation zone of the material, both springs experience a complete transformation from austenite to martensite state which is depicted in Fig. 3c. Here, the total martensite fraction for both springs starts from 1 and finally settles to 1 after going through a complete heating and cooling cycle. The net force exerted by the spring pair bends the tube. The maximum value of the bending angle for the given voltage pattern is found to be 22° in the positive direction and 11.3° in the negative direction, as indicated in Fig. 3d. Similarly, the results can be obtained for another antagonistic spring pair.

After estimating the motion of the flexible tube, the control algorithm is implemented through the LABVIEW platform. A square shape of 10 mm side length is drawn on the software and its vertices are provided as the desirable setpoints for the controller. Table 2 represents errors while tracing the desired setpoints. At first, the tip is made to reach the center point, and then depending upon the sequences of setpoints, the tip of the tube moves. While performing the control task, the sensor is continuously recording the data, thus a total of 1144 readings have been recorded. The desirable square shape and the tracked shape are shown in Fig. 4.

It can be seen from Fig. 4 that the manipulator follows the given square trajectory most of the time. As shown in Table 2, there is a slight off-track motion during tracing the (−10, 10) point, and a maximum error of −4.9 is observed on the *X*-axis during this time. Similarly, while tracing the (−10, −10) point, a maximum error in the *Y* axis is observed, which is −1.37 mm. The reason for such perplexing motion could be the unknown external disturbances that can be resolved by performing the recalibration task.

4 Conclusions

A working prototype of a flexible manipulator actuated by two pairs of antagonistic SMA springs has been mathematically modeled and simulated in a MATLAB environment. An existing loading history-based phase kinetics model has been adapted to establish the dynamics of an antagonistic SMA spring actuator-based flexible

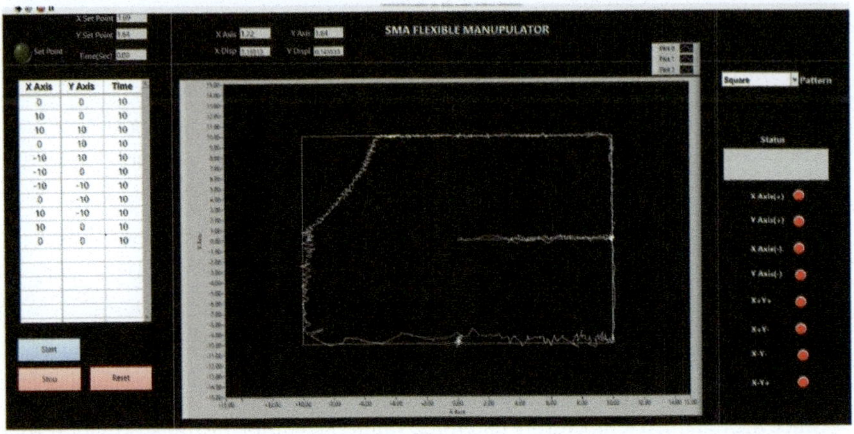

Fig. 4 Tracing of a square trajectory by the controller

manipulator. The simulation results predict a maximum of 6.5 N force that can be generated by the given antagonistic SMA spring actuator. This amount of force is sufficient enough to bend the tube by a maximum angle of 22°. Further, the tip of the flexible tube has been made to follow a given square trajectory via a calibration-based controller. The obtained experimental results show the trajectory tracing capability of the manipulator in 2D with a maximum positional error of 4.9 mm and 1.37 mm in the *X* and *Y* directions, respectively. Such SMA-based manipulators can be satisfactorily used in applications like endoscopy, tumor irradiation, etc. In the future, the authors will work on refining the control algorithm and tracing more 2D trajectories.

References

1. Garbin N, Wang L, Chandler JH, Obstein KL, Simaan N, Valdastri P (2018) A disposable continuum endoscope using piston-driven parallel bellow actuator. In: Proceedings of the international symposium on medical robotics (ISMR), Atlanta, GA, pp 1–6
2. Kellaris N, Venkata VG, Smith GM, Mitchell SK, Keplinger C (2018) Peano-hasel actuators: muscle-mimetic, electrohydraulic transducers that linearly contract on activation. Sci Robot 3(14)
3. Huang L, Liu B, Yin L, Zeng P, Yang Y (2021) Design and validation of a novel cable-driven hyper redundant robot based on decoupled joints. J Robot 2021:1–16
4. Liu J, Zhang C, Liu Z, Zhao R, An D, Wei Y, Wu Z, Yu J (2021) Design and analysis of a novel tendon-driven continuum robotic dolphin. Bioinspir Biomimet 16(6)
5. Choi J, Ahn SH, Cho KJ (2020) Design of fully soft actuator with double-helix tendon routing path for twisting motion. In: Proceedings of the 2020 IEEE/RSJ international conference on intelligent robots and systems (IROS), Las Vegas, NV, USA, pp 8661–8666
6. Sirohi R, Wang Y, Hollenberg S, Godage IS, Walker ID, Green KE(2019) Design and characterization of a novel, continuum-robot surface for the human environment. In: Proceedings of the 2019 IEEE 15th international conference on automation science and engineering (CASE), Vancouver, BC, Canada, pp 1169–1174

7. Hadi A, Akbari H, Tarvirdizadeh B, Alipour K (2016) Developing a novel continuum module actuated by shape memory alloys. Sens Actuators A 243:90–102
8. Tian J, Wang T, Fang X, Shi Z (2020) Design, fabrication and modeling analysis of a spiral support structure with superelastic Ni-Ti shape memory alloy for continuum robot. Smart Mater Struct 29:1–16
9. Hartl DJ, Lagoudas DC (2020) Shape memory alloys for aerospace, recent developments, and new applications: a short review. Proc IMechE Part G: J Aerosp Eng 221: 535–552
10. Nair VS, Nachimuthu R (2022) The role of NiTi shape memory alloys in quality of life improvement through medical advancements: a comprehensive review. J Eng Med 236(7)
11. Zhang ZX, Zhang J, Wu H, Ji Y, Kumar DD (2022) Iron-based shape memory alloys in construction: research, applications, and opportunities. Materials (Basel) 15(5):1–26
12. Ruth DJS, Sohn JW, Dhanalakshmi K, Choi SB (2022) Control aspects of shape memory alloys in robotics applications: a review over the last decade. Sensors (Basel) 22(13):1–17
13. Bhatt N, Soni S, Singla A (2022) Analyzing the effect of parametric variations on the performance of antagonistic SMA spring actuator. Mater Today Commun 31(103728):1–13
14. Banerjee A (2012) Simulation of shape memory alloy wire actuator behavior under arbitrary thermo-mechanical loading. Smart Mater Struct 21(125018):1–13
15. Bhatt N, Gurung H, Soni S, Singla A (2021) Effect of biasing conditions on the performance of an SMA spring actuator under thermo-mechanical loading. Mech Adv Mater Struct 1–15

Design and Control of All Pneumatic Virtual Motion Simulator

Ashish Siddharth, Arun Dayal Udai, and Sourabh Khemka

Abstract Virtual Motion Simulators are electro-mechanical systems that are used to create an immersive environment for virtual reality applications, like gaming platforms for the entertainment industry, flight training simulators, vehicle testing systems, etc. A Gough-Stewart parallel manipulator is one of the most widely used platforms to support these applications' human handling cockpits, along with its integrated controls. Most of these motion platforms were originally proposed with hydraulic jacks [7], which later on, with advancement in electrical controls, were reported with electrical actuators [8]. Hydraulic actuators find their application in robust and huge load capacity moving platforms, whereas electrical actuators are applied mainly for applications with high stiffness and precision, like telescope mounts [14]. Both hydraulic and electrical actuators pose a trade-off of being expensive for non-critical applications like gaming platforms or training simulators where a high degree of precision is not much of a concern. On the contrary, pneumatic actuators are inexpensive, fast and offer a long linear actuation. Reckoning this, we propose using pneumatic piston linear actuators for all six legs of a Gough-Stewart platform for a Virtual Motion Simulator. However, due to the compressibility of air in the linear cylinder, the actuators are prone to external disturbances in their position due to the dynamically varying load while the platform is moving. The paper proposes to use Continuous Integral Sliding Mode Control for the precise positioning of six pneumatic actuators of the Virtual Motion Simulator. The first part of the paper will discuss the mechanical design, the kinematics and physical assembly of the motion platform. Secondly, the paper will discuss the hardware integration of the electro-pneumatic system, which implements the controller for all the six linear actuators of the Gough-Stewart platform. Next, the paper will discuss the control

A. Siddharth · A. D. Udai (✉)
Indian Institute of Technology, Dhanbad, India
e-mail: arunudai@iitism.ac.in

A. Siddharth
e-mail: asiddharth.20dr0175@mech.iitism.ac.in

S. Khemka
Birla Institute of Technology, Mesra, India

S. K. Saha and K. Rama Krishna (eds.), *Recent Developments in Multibody Dynamics*, Lecture Notes in Mechanical Engineering,
https://doi.org/10.1007/978-981-97-7525-5_4

approaches for precise positioning of the pneumatic actuator, namely, the continuous integral sliding mode control technique. Finally, the hardware integration to the PC-based gaming engine is discussed, and the methods for extracting the data, pose, velocity and acceleration from the gaming engine are discussed. The paper will conclude with results and a discussion on the performance of the motion simulator as a whole.

Keywords Virtual motion simulator · Parallel platform · Sliding mode control · Electro-pneumatics

1 Introduction

The parallel robot was introduced concurrently with serial robots in the year 1954 [9]. Parallel robot owns various advantages such as high power to weight ratio, high payload capacity (applied in telescope mounts, simulators and machine tool handling), high precision (required in the electronic industry, medical surgery) and high speed (a necessity for food processing, pelletizing) and small footprint (profitable in 3D Printers, mechanical stabilisers). However, even today, parallel robots are sparse in the industry due to complex physics [12, 27], intricate mechanical structure [20, 22, 25], and demanding hardware and software controller requirements. Additionally, due to the difficulties described above, it became less visible in academic curriculum and robotics laboratories across the country. Few robot manufacturers, like ABB, $FANUC$, $ADEPT$ and a few others, came forward to address the industrial requirement that can only be resolved precisely by parallel robots. This resulted in the exorbitant pricing of these robots. Also, the most popular types these manufacturers make are of three degrees of freedom (DoF).

Parallel six-DoF manipulators currently available in the market are $ACROME$ and PI Hexapod. In India, $Hexa$ [29] is a commercially available parallel platform designed and developed at $ADUZA$. Various others are reported by Dasgupta [5] and Milica [24]. Both systems are of Universal—Prismatic—Spherical (UPS) structure, acquiring design similarity with the Gough–Stewart platform [31]. This type of mechanical structure later found its way in the development of virtual motion simulators. Motion simulators are frequently utilised for instructive and educational purposes. These systems can produce a virtual driving condition that can be utilised to prepare freshmen before entering the real world. Besides that, motion simulators are significant in information assortment for road well-being research, vehicle framework development, human behaviour study and traffic control gadget advancement. These permit researchers, developers and ergonomists to plan the design according to human comfort. This also improves the procedure for vehicle planning.

The first known driving simulator was a joint research of General Motors with Virginia Polytechnic Institute and State University in 1970 [1]. This simulator was mainly developed to study the effect of interior design on driver performance. The simulator had two degrees of freedom with the capability to roll and pitch on its axis.

Later in 1983, the federal highway administration driving simulator (HYSIM) started working on driver reactions to different driving experiences for specific studies. Their study includes a human response to the different background colours of stop signs, signals, designs and other warning signs in various working zones [28]. Later in 2011, the Swedish National Road and Transport Research Institute (VTI) [16] developed such a simulator using a hydraulic actuator. They have been working on developing such a simulator since 1984. The current model is a fourth advanced driving simulator built with a roughly 2.4 million euros budget. Few universities, such as San Jose State University, successfully built a gaming simulator using a motion platform actuated by all rotary electric motors. Their motion platform could only simulate the roll and pitch motions [6]. Loughborough University has reported works on Stewart platform-based driving simulator and fabricated a pneumatic operated Stewart platform for an industry-sponsored project. However, it could only perform basic motions and was developed for a small payload [11]. The University of Stuttgart also built a pneumatically driven parallel platform using motion and force control for flight simulator applications. However, they have used an additional cylinder in the centre, which restricts the other three degrees of freedom [26].

High-payload parallel platforms are primarily built using hydraulic actuators. However, such a platform is quite expensive, has higher stiffness and poses safety issues for human handling [37]. Having electrical actuators is reasonably less expensive and precise as well. The technology for driving linear electrical actuators makes it inherently slow and limits its applicability in faster gaming platforms. This paper proposes a virtual gaming six-DoF simulator (having three linear movements: linear X, longitudinal Y, vertical Z and three rotations movements, pitch P, roll R and yaw Y) using pneumatic actuators. An electronic circuit for digital motion control of the platform is also developed. These pneumatic actuators are inexpensive, safe and reasonably faster than electrical or hydraulic actuators. The pneumatic actuator may have less accuracy with the tested controls [13]. However, this would hardly create problems with commercial gaming platforms for the proposed scale of 300 Kg payload capacity. This pneumatically controlled virtual simulator will be the first of its kind in India to be used for commercial gaming platforms.

2 Design and Assembly

The design of parallel mechanisms is an important and challenging problem in robotics. The most important issue while designing such a manipulator arises while solving its kinematic equation, checking its quality index and performing its performance evaluation [19]. The other important issue is determining the manipulator's dimensions, which are most suitable for the required task. The performance of a parallel manipulator heavily depends upon its dimensions [18]. Hence, choosing the best kind of system for a required application is a significant challenge. Therefore, designing a manipulator is mainly dominated by: workspace, dexterity, payload and type of task it will perform. In this section, the design of a parallel platform has been described, and further, the model has been tested to study the position control with the electro-pneumatic actuator.

2.1 Design Process and Specification

The complete design process and specification can be incorporated by subdividing it into three stages. These three design stages are described as follows: 1. Product Specification and Planning Stage. 2. Conceptual Design Stage. 3. Physical Design Stage [32].

2.1.1 Product Specification and Planning Stage

In this first stage, dimension, functionality and other requirements are specified for the product development. The geometrical and functional requirements of the product are specified as follows: 1. The mechanism must be a spatial parallel structure. 2. The mechanism must have six degrees of freedom. 3. The mechanism should be capable to move freely within the workspace without any hindrance caused due to any bearing constraints 4. The mechanism must have the ability to carry 300 Kgs of load. 5. The structure should be compact. Since the Stewart platform developed employs a UPS-type mechanism, it inevitably satisfies the first two points of our product criteria. Each leg in this mechanism, having prismatic and spherical joints, makes it in compliance with criteria point number three and four. The letter U, P, S, R are used to indicate Universal, Prismatic, Spherical and Rotary actuators or joints, respectively. Various combinations may be achieved depending on the type of applications. Some example are UPS, RPS, RUS, RSS, etc. Hence, choosing the best among these systems for a required application is a major challenge. Generally, for a system with real-time feedback and with faster response, a six UPS structure is considered. UPS structure has improved efficiency in computational kinematics, dynamics, singularity and optimisation [21]. One of the significant advantages of UPS mechanism is that it can withstand a large payload with relative compact sizes compared to other serial or parallel geometries. Hence, the dimensional requirement for a compact size structure is met during stages of the design process.

2.1.2 Conceptual Design Stage

The conceptual design stage is the second stage, where various rough conceptual design alternatives are developed according to the product specification, and one with the best performance is chosen. The parallel octahedral structure was the first considered since it is the simplest structure of its kind [15]. As per Grubler criterion [32], the mobility of general spatial mechanism is given by:

$$M = \lambda(n-1) - \sum_{i=1}^{j}(\lambda - f_i) \tag{1}$$

where:
M: Mobility, or number of degree of freedom of the system.
λ : Degrees of freedom of the space in which a mechanism is intended to function, for the spatial case, $\lambda = 6$.
n: Number of links in the mechanism, including the fixed link.
j: Numbers of joints in a mechanism, assuming that all the joints are binary.
f_i : Degrees of relative motion permitted by joint i.

The six-six octahedral parallel structure shown in Fig. 1 has a top platform, a base platform and a six-leg connector with spherical joints and universal joints. Each leg connector is treated as two bodies connected by a prismatic joint. Hence, there are 14 such bodies: the top platform, the base platform and two for each of the six legs. Each universal joint has two relative degrees of freedom, and the spherical joint has three relative degrees of freedom. There are 12 such joints, with 6 on the top platform having spherical joints and the other 6 on the base platform having a universal joint. The legs numbered one to six are arranged to assist each leg in being connected either of the corner points of the base platform or the top platform.

In order to calculate the mobility of the above six-six parallel platforms, Eq. 1 is used:

$$M = 6(14-1) - \sum_{i=1}^{6}(6-3) - \sum_{i=1}^{6}(6-2) - \sum_{i=1}^{6}(6-1)$$

$$M = 78 - 18 - 24 - 30 = 6$$

Hence, six degrees of freedom are found for the chosen structure, which satisfies our criteria.

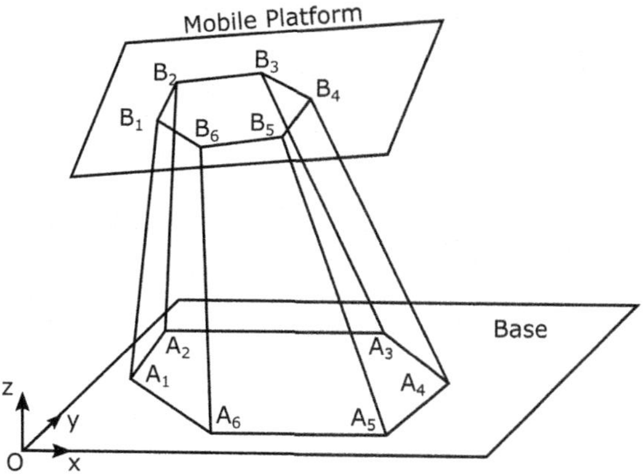

Fig. 1 Kinematic structure of six-six octahedral parallel manipulator

2.1.3 Prototype Design

Once the desired configuration was finalised and a rough structure was ready. More detailed and specific information about the parallel platform for UPS type mechanisms as discussed in the conceptual stage is decided. The entire design of the Gough-Stewart platform involved theoretical and practical considerations. The idea was to build a driving simulator where the top platform could execute six degrees of freedom. Since a driving simulator is one of the Stewart platform's applications, choosing the UPS type mechanism was an ideal choice [21]. Also, the structure should be designed in a way such that it can handle the human load, keeping safety standards and ergonomics rules in top priority. Hence, the design was proposed to bear a load of 300 Kg without any displacements or signs of failure.

Finally, at the last stage, the design is dimensionally verified as per manufacturing standards and is sent for analysis and optimisation. Once analysis and optimisation are over, the product is sent for finite element simulation and accordingly, the material selection, shape and size are decided considering factors like safety, ergonomics and aesthetics. After these drawings are prepared, and it is sent for manufacturing. Although the final stage is generally time-consuming, the first two stages significantly impact cost control and design results.

2.1.4 Design of Parallel Structure

For the given payload criterion, $CE1D63 - 500$ stroke reading pneumatic cylinder manufactured by SMC India private limited was selected. The cylinder has a stroke length of 500 mm. The length of the body of the cylinder is 645 mm. The bore diameter of the cylinder is around 63 mm, capable of holding maximum and minimum operating pressure of 1 MPa and 0.05 MPa, respectively, according to product specifications. The cylinder home position can be set anywhere within the cylinder stroke. When the cylinder is reset, the cylinder's current rod position becomes the home position. Initially, the middle position of the cylinder was considered starting home position to get free motion without any interference. Total leg length was later calculated by summing up the length of a pneumatic cylinder, the axial length of spherical joints, UV joint till its centre of location and extension due to wedges on the chassis. The calculated length from the CAD software was found to be around 950 mm.

Considering feasibility requirements, about dimensional limits of commercial and manufactured parts available in the market and the required working space present at the installation site, a CAD $3D$ model was created.

2.2 Design of Cockpit

Anthropometry is a branch of ergonomics used to set up an environment for a workstation design to optimise human interaction. The ultimate goal of these design principles is to enhance human performance, control fatigue and prevent accidents [38]. While designing the cockpit, specific considerations have been incorporated to have flexibility in the shape and size of the user operating the cockpit. These considerations include human factors, user population, body measurements and more. The design has been made in accordance with $FSAE\ SAE$ rule $A1.2.2$ [23], which suggests the vehicle cockpit should be such that it can accommodate drivers whose stature value varies between 5th percentile female and 95th percentile male. The guidelines also mention that if the design is per anthropometric measurement, then the stature takes care of driver visibility, steering wheel, shiftier and pedal mounting locations, belt angles and head-space of the designed cockpit. Hence, fulfilling the requirement for cockpit design. Also, the proposed design considers anthropomorphic data across various countries to have the desired user flexibility, cover a broader potential and market, have a wider reach and hence be scalable.

For this project, three HP Omen $4K$ screens were mounted on the platform cockpit and have been set according to drivers' visibility. Considering safety as a top priority, the emergency stop button has been provided in three locations. It has been placed in positions where instant human reflexes go in case of emergency. Two emergency stop buttons were placed near the steering wheel and one on the side of the seat and gear shifter. There were two extra emergency buttons, one near the operator's cabinet and one at the programmer's end. These buttons locked the top platform in the same pose when pressed. Also, arrangement for driver safety belt has been provided, and proper safety markings and safe operating procedure (SOP) have been made per industrial standards. Arrangement for signals and sound systems has also been developed. When the system starts, a hooting sound occurs, followed by a yellow and green signal. The green signal gives the driver visual information to start running the simulator. During standby mode, a red visual signal and a similar hoot sound are heard when the system goes into standby mode. The Digital 5.1 Dolby sound system was also set up to provide the feeling of engine vrooming when the gas pedals are pressed while the simulator is running. Users are also given the option to choose between manual or automatic gear shifting mode. The driver can choose between six different gear modes in manual mode, providing the ultimate real-time experience while running the simulator. The final CAD model is shown in Fig. 2.

3 Hardware Setup and Integration

This section explains the hardware setup required for position-controlling pneumatic cylinders. Figure 3 shows the initial setup used to test the position control of the pneumatic system. The system controls the position of a $SMC\ CE1B40 - 100$

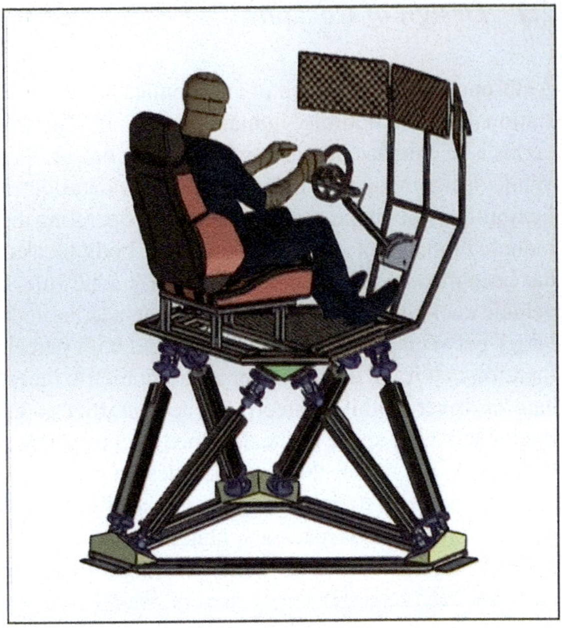

Fig. 2 CAD model of complete structure

(stroke reading cylinder) using electro-pneumatic controller. The position feedback pulses were fed to the SMC multi-counter ($CEU5P - D$) and were directed to the controller through the $RS - 232C$ to USB converter. The analogue output for commanded signals was generated using a custom-designed human interface device (HID) class USB based data acquisition system (DAQ) that uses $ATMEL$ $AT90USB1287$ micro-controller.

The analogue output generated by the DAQ board is of eight-bit type. The change in length obtained through solving inverse kinematics for a given pose is mapped with this eight-bit signal [34]. These signals are then fed to a power amplifier $VEA252$, which controls the current flowing through the solenoids of an electro-pneumatic proportional flow control valve $VEF312 - 1$. The solenoids of open the valve's poppet as per the amount of current flowing through it. Two such valves control the flow of air on either side of the double-acting cylinder. Airflow from these two ports determines the position of the piston inside the cylinder. When the flow of air is increased from one port and decreased from another, the piston moves to obtain its new equilibrium position. Thus, the difference in air flow rate between these two valves' piston positions can be controlled. The suitable control algorithm for incorporating such system is discussed in next section. Considering safety at top most priority, emergency lockup valve ($IL211$) can be activated in case of unavoidable situation to constrain the position of the cylinder. Figure 4 shows the control flowchart for the above system.

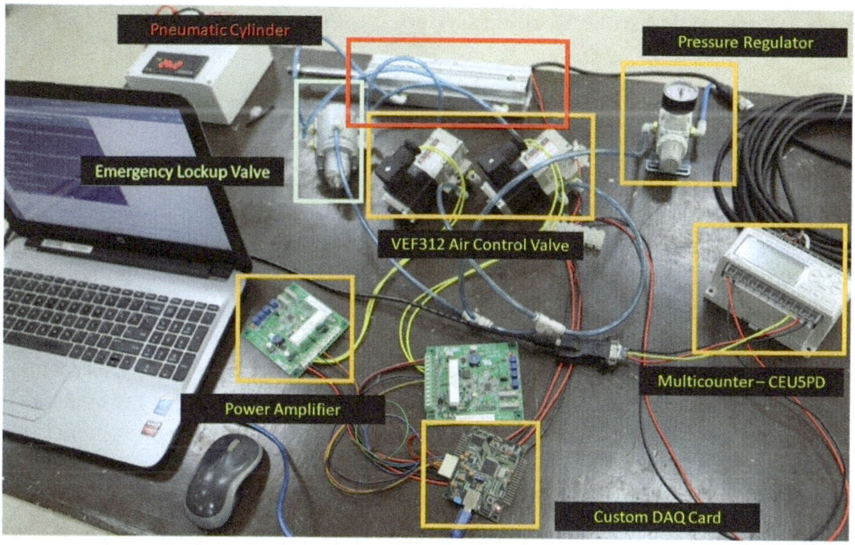

Fig. 3 Control of a single SMC stroke reading cylinder

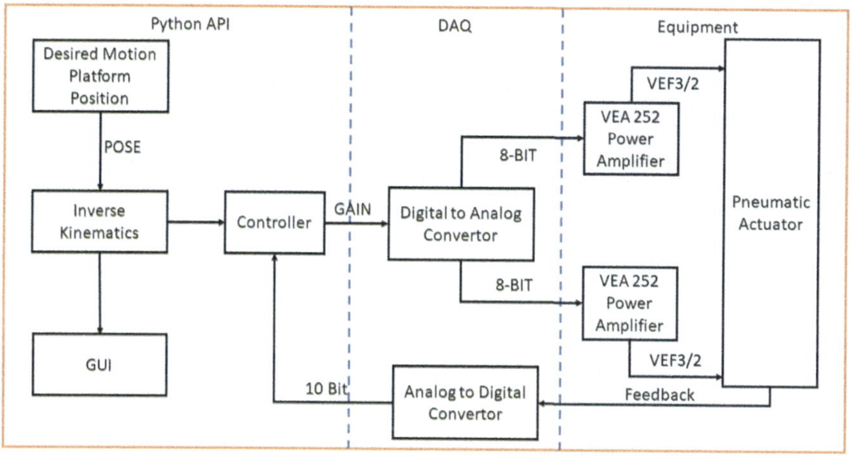

Fig. 4 Flowchart displaying control of a SMC stroke reading cylinder

3.1 Electro-pneumatic Circuit Components

This section explains the components used to perform position control of the pneumatic cylinders. The working of these specific components, along with their working principle, has been discussed below.

3.1.1 Pneumatic Stroke Reading Cylinder

The proposed design has used the SMC pneumatic stroke reading cylinder with model number CE1B63-500. The model itself has been manufactured by SMC pneumatic India private limited. This pneumatic cylinder is double acting, and the rod is a non-rotating type with a resolution of 0.1 mm. The model number describes the features of the cylinder. The first three letters describe the base part number to which it belongs. B stands for direct mount type structure, and 63 mm is the bore diameter of the cylinder. The stroke length of this cylinder is 500 mm, and the maximum operating pressure of this cylindrical model is 1.0 MPa.

Measuring Principle: The scale present in the piston rod consists of a magnetic layer and a non-magnetic layer in 0.8 mm pitch. The magnetic resistive element, as shown in Fig. 5, notices a signal in two phases in the form of *cosine* and *sine* due to the movement of the piston rod. The waveform generated due to the movement of the piston rod has a pitch of about 0.8 mm in a single cycle. This signal is amplified eight times, and values are further divided by eight to get actual measurements. Doing so, a 90° phase difference is observed in signal two, having a pulse of 0.1 mm.

3.1.2 Multi-counter $CEU5P - D$

The multi-counter count pulses sent from the encoder attached to the actuator in the form of A/B quadrature pulse. It then sends the signal to the controller, where the control algorithm works. The feedback is sent via $RS - 232$ serial communication protocol. $RS - 232C$ standard is most widely used and supported by computers and micro-controllers. It eliminates the clock signal wire; hence, asynchronous transfer mode was observed.

Receive (RX) and transmit (TX) for master and slave devices are cross-connected, and each may drive the other. Each sends and receives the data independently. As there is no clock signal, the data transfer speed is predetermined by a baud rate. The number of bits transferred per second is given by the baud rate. This serial data is

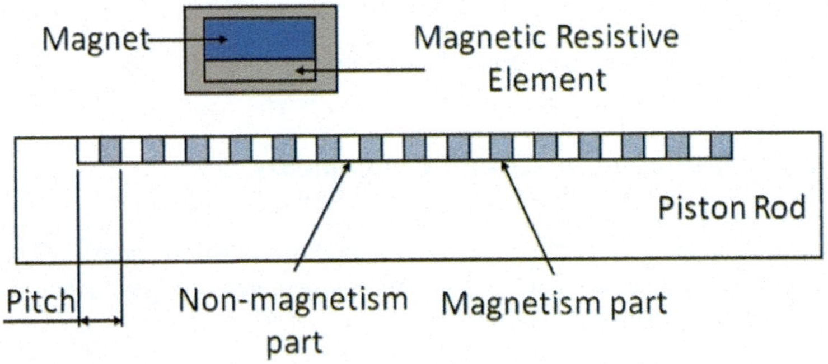

Fig. 5 Pneumatic stroke cylinder measurement principle

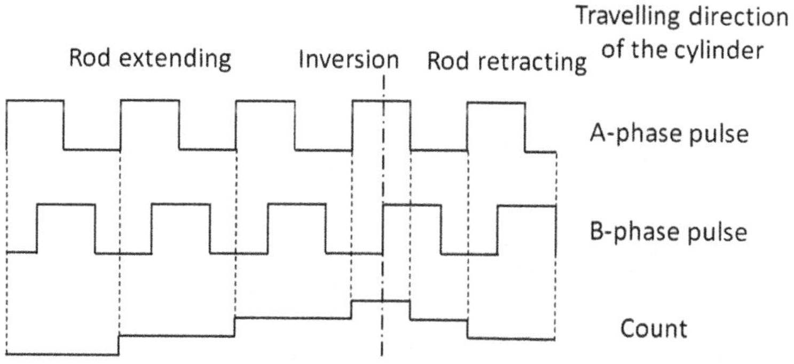

Fig. 6 Relationship between pulse and count

formatted sequentially as a start bit, eight data bits, one parity bit and one or two stop bits.

Relationship between pulse and count: Since the magnetic resistive element receives a signal in two phases, so when the piston rod is moving out, the pulse is received from the magnetic region first and then from the non-magnetic region, with a phase difference of 90°. Similarly, when the piston rod is moving in, the pulse from the non-magnetic region comes first, followed by the magnetic region with a phase difference of 90°, as shown in Fig. 6. The multi-counter interprets these pulses and performs addition in the first case and subtraction in the other.

3.1.3 Eight Bit, 12-Channel Data Acquisition System

This custom-designed eight-bit, 12-channel analog output board belongs to the HID class device using USB based $ATMEL$ $AT90USB1287$ micro-controller. The board has 12 channel outputs to control the flow rate of twelve different electro-pneumatic valves. One double-acting pneumatic cylinder requires two electro-pneumatic valves to control the position. Since there are six actuators, 12 output channels were required. The voltage output from these 12 channels is in the $0 - 5V$ range. The provision for four digital outputs was also made to handle the status of signal lights during the actual implementation of this data acquisition system.

3.1.4 Air Compressor

Since the working of the parallel structure depends entirely on pressurised airflow, and primarily pneumatic components have been used, an external compressor was needed to hold the compressed air. For the proposed structure, a 5 horsepower 80 litres capacity air compressor was used as it was found to be sufficient to drive six pneumatic actuators. The pressure values in each leg of the platform at any instant are different.

3.1.5 Lockup Valve

For safety, a lockup valve was included in the pneumatic circuit. In unavoidable circumstances, when an emergency push button is pressed, the lock-up valve locks up the air supply inside the pneumatic cylinder, keeping the pneumatic piston in a stable position. The emergency lock-up valve with model number $IL211$ manufactured by SMC (India) was used for the proposed structure.

3.1.6 Power Amplifier

The power amplifier VEA 252 manufactured by SMC was used in this work. It was needed to provide the processed signal received from the data acquisition system (DAQ). Since the required current from DAQ does not have enough power to control the proportional valve, an external current source amplifier is required.

3.1.7 Flow Control Valve

The flow control valve is needed to maintain flow at a different section of the pneumatic cylinder. The flow control valve has a proportional solenoid attached to it. The solenoid controls the puppet position inside the valve, hence maintaining the output flow. The flow control valve manufactured by SMC India, having model number VEF 3121, was used for this purpose.

4 Control Algorithm

Controlling any system is defined as the ability of the system to behave as per user commands or instructions. The actuator would not move as per the commanded trajectories in a real scenario due to external forces like friction, viscosity, backlash and many more. Hence, the deviations from the actual positions are sensed and fed back to the control system to generate corrective forces to minimise the error. In robotics, a manipulator is a physical system that needs to be controlled. There are several ways to implement a control algorithm and the way they are implemented significantly affects system performance. Also, the design of mechanical structure and type of actuator used are vital in choosing the type of control needed. The mechanical structure significantly influences controller design as it affects the structure stability, whereas the type of actuator used affects the control strategy. This paper uses

Proportional Integral and Derivative (PID) controller and Continuous Integral Sliding Mode Control (CISMC) with Linear Quadratic Redulator (LQR) to stabilise the pneumatic pistons at desired stroke length [17].

4.1 PID Controller

PID controller is a second-order linear control model and is a combination of three different controllers viz P, I and D. This controller is widely accepted in the industry when it is difficult to find system parameters. The transfer function for any general form of PID Controller is shown in Eq. 2.

$$C(s) = K_p + K_i/s + K_d s, \tag{2}$$

where K_p is the proportional gain, K_i is the integral gain, K_d is the derivative gain and s is the Laplace variable. The gain values are selected after rigorous tuning by varying controller gains [39]. The selection of proportional gain with gain K_p will decrease rise time and steady-state error. Whereas derivative gain with gain K_d will make the system more stable by decreasing overshoot and improving transient response. Finally, selecting integral gain would eliminate steady-state error but may worsen transient response. Further, all these gains are correlated; changing one may affect the other two.

The change in length obtained from computational model is actual input, which is fed in PID control system. The control system then calculates error from the feedback and sends the control gain signal to pneumatic flow control valve through the power amplifier.

4.2 Sliding Mode Controller

Pneumatic systems are nonlinear in nature, as it uses air as a fluid medium to drive themselves [35]. Air is highly compressible. Hence, a substantial difference or inaccuracy between input and output signals is observed for a dynamic system [2]. The PID controller fails to maintain positional accuracy for such a dynamic system. The system oscillates if tuning is not done correctly and is hazardous for a pneumatic system [36]. Hence, a better controller such as sliding mode control is needed to prevent such issues for a pneumatic system [10]. The sliding mode system drives the framework inside a continuous switching manifold at a high frequency. This switching manifold is called a sliding surface, where the framework slides or converges to its origin point asymptotically. Since the control input is at high frequency, jittering effects are observed inside a mechanical system, limiting its practical applicability. Hence, a modified form of sliding mode control that is a continuous integral sliding mode (CISMC) along with a super twisting algorithm has been reported here, which causes the system to slide to its origin without any disturbance [4].

Control input for CISMC is combination of super twisting control and a nominal control,

$$U = U_{STC} + u_{nominal} \tag{3}$$

where,

$$U_{STC} = -k_4|S|^{\frac{1}{2}} sign(S) + v \tag{4}$$

$$\dot{v} = -k_5 sign(S) \tag{5}$$

$$S = x_n - x_{n_0} - \int_0^t u_{nominal} d\tau \tag{6}$$

Here, x_n is n^{th} state of the system, S is the sliding surface, k_4 and k_5 are the gains for super twisting algorithm. U_{STC} creates a disturbance-free environment for controllers like PID or LQR. So, when the control input U is applied to the system, it stabilises the system states at desired points preventing disturbances.

An LQR control system requires information on all the states of the system. The system states are acquired either by having a dedicated sensor unit or by estimating a few using mathematical operations like differentiation. The pneumatic system using sliding mode along with LQR control, as shown in Fig. 7, has feedback for position and pressure. The piston's velocity is calculated by taking the time derivative of position. Position values are recorded with the help of magnetic position sensors in-built into the cylinder rod, and pressure is recorded using SMC built $PSE530-M5-L$ pressure sensors. These state values are input into a Python script which implements the sliding mode algorithm.

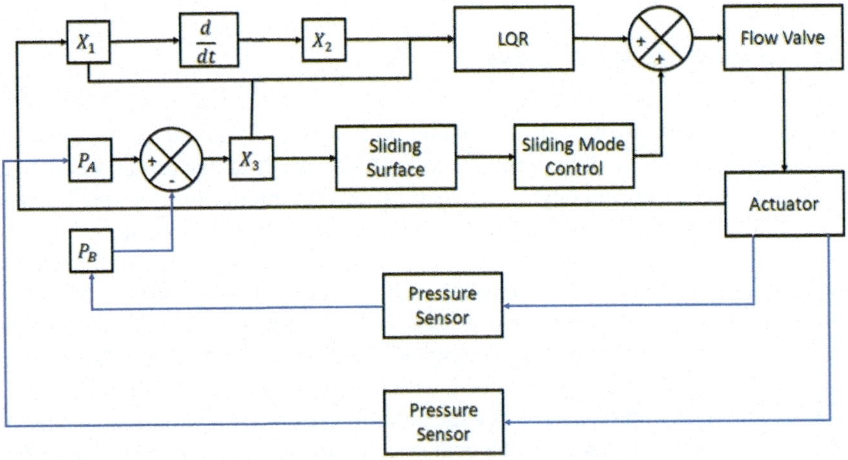

Fig. 7 Flowchart of sliding mode control

5 Results and Discussion

The results obtained from simulation and experimentation are described below.

5.1 *Inverse Kinematics Simulation Result*

The UPS parallel structure's graphical representation can be found on the website [30]. The graphical user interface (GUI) has been developed using WebGL Technology. Threejs [3] library has been employed to create the simulated environment. The GUI gives a visual representation of the top platform by changing pose and orientation. Cartesian control of the parallel manipulator can be provided as input on the left-hand side of the GUI. On giving motion to the top platform, the inverse kinematics function is called, which returns the required link length. The change in link length is displayed at the top header of the GUI.

Simulation for the motion platform was also performed in MATLAB. The inverse kinematic function was created in MATLAB which when called upon gives the required link length for the prismatic joint. Figure 8 shows the stick line diagram of the base platform and top platform at home and at position: $x = 10$ mm, $y = 15$ mm, $z = 20$ mm, $r = 10°$, $p = 15°$, $y = 20°$.

5.2 *PID Controller Result*

Initially, the output of each pneumatic cylinder's was recorded and compared with the input signal. This was done as the weight experienced by each cylinder was different.

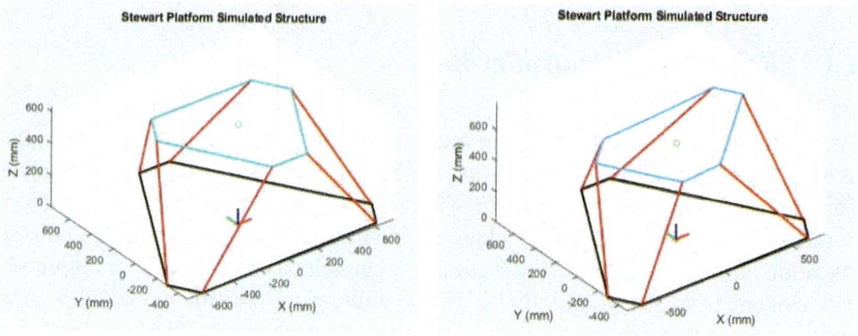

(a) Pose1 (b) Pose2

Fig. 8 Plot in MATLAB for home pose and pose (10,15,20,10,15,20)

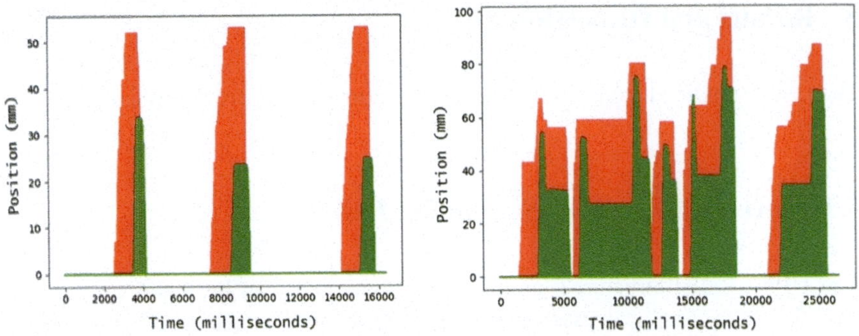

(a) Plot 1

(b) Plot 2

Fig. 9 Final output plots with PID control algorithm

Hence, the gain values required in the control algorithm will be different. Proper control gain values were decided onsite after running feasibility tests in large numbers. Figure 9 shows the variation in output with the command signal sent. The plot with Red Colour represents the actual position input given while the one in green color represents the final output received via feedback. There is a margin of error which varies for different positions which may be due to the high compressibility of air or the stick-slip behaviour or jerking motion of the cylinder. Hence, exact positioning becomes very difficult when there is a dynamic change in the system. However, on the positive side, there is no oscillation, as seen in the plot; the system behaved positively and followed the critical path as per commanded signal with a margin of error. Also, the pneumatic cylinders are overpowered for the given requirement, which avoids oscillations under any commanded position. Hence, purely positioning control is achieved as the overpowered actuators are hardly affected by the system's dynamics.

5.3 *Sliding Mode Controller Result*

The simulation results of CISMC are presented in Fig. 10. A finite time stabilisation of the system states is achieved despite the high amount of nonlinear disturbances. The reference position state as shown in Fig. 10a is changed from 100 mm to 90 mm. Figure 10b shows that the piston velocity is stabilised to a small region around zero. Once the position is stabilised, variation seen in velocity is very low in amplitude. It is evident from the graphs that it is maintaining its position set-point without any momentary significant glitches in the velocity-time graph. The low amplitude variation present in velocity will be felt only as in-significant vibrations.

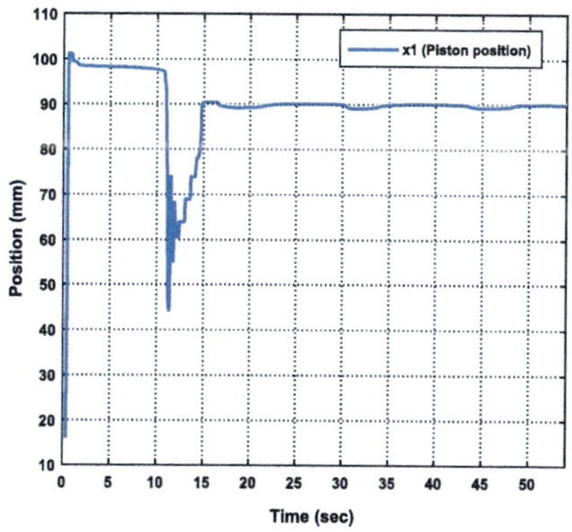

(a) Plot 1

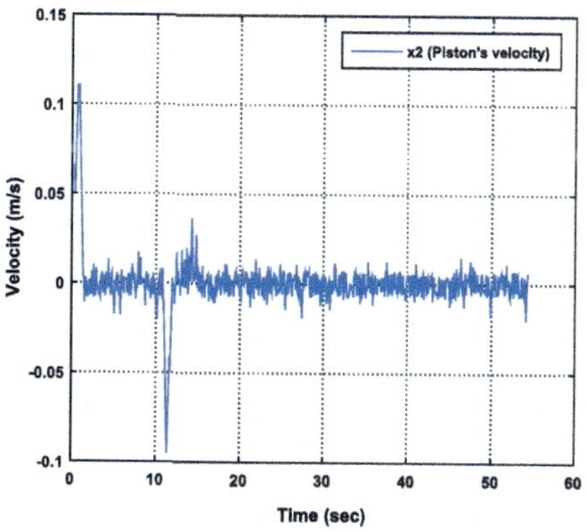

(b) Plot 2

Fig. 10 Final output plots with sliding mode control algorithm

5.4 *Virtual Motion Simulator Final Assembled Design*

The representative Fig. 11 shows the indigenous developed view of the platform designed with pneumatic actuators. The required roll and pitch angle of $+/-25$ degrees was achieved through this design. The system behaved positively for any given trajectory required for motion simulation. The pose values were obtained through a live game engine of Project Cars2 and the top platform was seen to be oriented as per game frame. As the platform has limited displacement capabilities along any direction, making the rider feel the acceleration using real translation rates would not last long. The accelerations are simulated to the rider using the tilt angles of the platform and vary within $+/- g$. All the accelerations are simulated using tilt along X or Y directions. Hence, the driver can feel the tilt and pitch of the game while running it. The final working demo video can be found on the YouTube channel [33].

6 Discussion

Most gaming platforms are electrically actuated, expensive and mostly made through expensive imported linear actuators. The novelty here is the use of stroke-reading pneumatic cylinders, which are far more inexpensive than the electrical actuators of similar thrust. The realisation of pneumatics for a 300 Kg payload Stewart platform-based gaming platform/VR simulator is being reported for the first time. The contribution and novelty of this research are:

- The complete research project is an initial endeavour to control the motion of the mobile platform.
- The design of complete platform was developed indigenously from scratch.

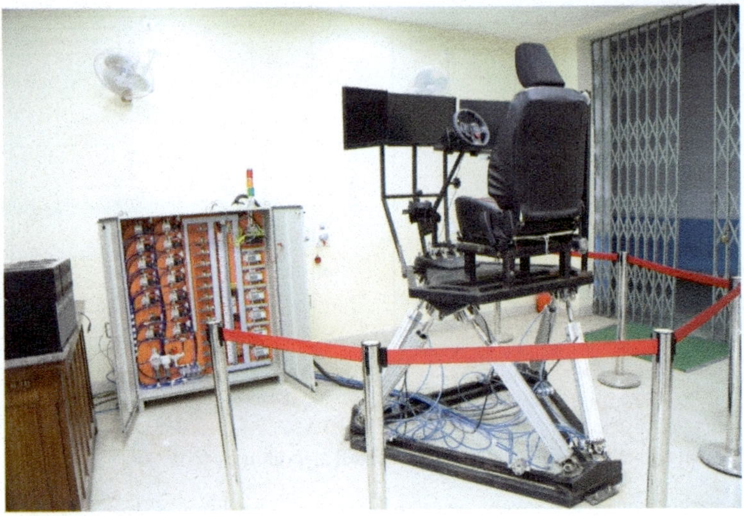

Fig. 11 Developed Stewart platform for motion simulation

- Use of any kind software was avoided to run this platform. The motion platform works independently, all codes were written in Python. APIs in Python were developed to integrate the system with hardware.
- The data acquisition system was also indigenously developed to control the actuator position.
- The control algorithm developed was simulated in custom developed graphics software and tuning for the same was done after rigorous testing.
- A graphical simulation for the same was also developed using JavaScript [30] to understand the motion of platform. The use of computer applications was limited to avoid the black-box in development process.

7 Conclusion

This study proposed the development of a virtual motion platform using pneumatic actuators. The platform was designed, developed and successfully tested [33]. The platform developed can be implemented as a driving test system as it provides a controlled domain wherein driver reactions to traffic control can be seen without danger. Also, the potential danger can be avoided in a short measure of time. One of the more obvious points of interest is that a driving test system doesn't open members to threats. Auto collisions are consistently a potential threat in the field, mainly when human interaction is involved; be that as it may, there are no mishaps in the simulation system. Motion platforms can also be leveraged to test the new roadways in the design stage without building roadway costs.

Based on the performance of the developed motion platform, it has the potential to be employed at a large scale and can be commercialised.

Acknowledgements The author would like to thank Mr. Ritanshu, Mr. Aryendra Singh, Mr. Ankit Kumar and Mr. Arnav Kumar for their constant support, constructive criticism and encouragement in this project.

References

1. Bertollini G, Johnston C, Kuiper J, Kukula J (1994) The general motors driving simulator. SAE Technical Paper 940179. https://doi.org/10.4271/940179
2. Bone GM, Ning S (2007) Experimental comparison of position tracking control algorithms for pneumatic cylinder actuators. IEEE/ASME Trans Mechatron 12:557–561. https://doi.org/10.1109/TMECH.2007.905718
3. Cabello R, Brunt P, Ulicny B, Koo J (2014) Threejs library website. http://threejs.org. Accessed June 2014
4. Chalanga A, Kamal S, Bandyopadhyay B (2015) A new algorithm for continuous sliding mode control with implementation to industrial emulator setup. IEEE/ASME Trans Mechatron 20:2194–2204. https://doi.org/10.1109/TMECH.2014.2368717
5. Dasgupta B, Mruthyunjaya T (2000) The Stewart platform manipulator: a review. Mech Mach Theory 35:15–40. https://doi.org/10.1016/S0094-114X(99)00006-3
6. Dynamics FM (2018) San Jose University. https://www.youtube.com/watch?v=zNUBZfrOXUc&ab_channel=FullMotionDynamics. Accessed 15 July 2018
7. Gallardo-Alvarado J (2016) Gough's tyre testing machine. In: Kinematic analysis of parallel manipulators by algebraic screw theory. Springer International Publishing. https://doi.org/10.1007/978-3-319-31126-5_12
8. Ghobakhloo A, Eghtesad M, Azadi M (2006) Position control of a Stewart–Gough platform using inverse dynamics method with full dynamics. In: 9th IEEE international workshop on advanced motion control, pp 50–55. https://doi.org/10.1109/AMC.2006.1631631
9. Gough VE (1956) Contribution to discussion of papers on research in automobile stability, control and tyre performance. Proc Auto Div Instit Mech Eng:392–394
10. Grewal KS, Dixon R, Pearson J (2011) Control design for a pneumatically actuated parallel link manipulator. In: 21st International conference on systems engineering, pp 43–48. https://doi.org/10.1109/ICSEng.2011.16
11. Grewal KS, Dixon R, Pearson JT (2012) LQG controller design applied to a pneumatic Stewart–Gough platform. Int J Autom Comput 9:45–53. https://doi.org/10.1007/S11633-012-0615-7
12. Griffis M, Duffy J (1989) A forward displacement analysis of a class of Stewart platforms. J Robot Syst 6(6):703–720
13. Hildebrandt A, Neumann R, Sawodny O (2010) Optimal system design of SISO-servopneumatic positioning drives. IEEE Trans Control Syst Technol 18:35–44. https://doi.org/10.1109/TCST.2008.2009879
14. Ho PTP, Altamirano P, Chang CH, Chang SH, Chang SW, Chen CC, Chen KJ, Chen MT, Han CC, Ho WM, Huang YDT, Hwang YJ, Ibañez-Romano F, Jiang H, Koch PM, Kubo DY, Li CT, Lim J, Lin KY, chin Liu G, Lo KY, Ma CJ, Martin RN, Martin-Cocher PL, Molnar SM, wang Ng K, Nishioka H, O'Connell KE, Oshiro P, Patt F, Raffin P, Umetsu K, Wei T, Wu JHP, Chiueh T, Chiueh T, Chu TH, Huang CWL, Hwang WYP, Liao YW, Lien CH, Wang FC, Wang HF, Wei RM, Yang CH, Kesteven MJ, Kingsley JS, Sinclair MM, Wilson WE, Birkinshaw M, Liang H, Lancaster K, Park CG, Li Pen U, Peterson JB (2014) The Yuan-Tseh Lee array

for microwave background anisotropy. Astrophys J 694:1610–1618. https://doi.org/10.1088/0004-637X/694/2/1610
15. Hunt KH, McAree PR (1998) The octahedral manipulator: geometry and mobility. Int J Robot Res 17(8):868–885
16. Jansson J, Sandin J, Augusto B, Fischer M, Blissing B, Källgren L (2014) Design and performance of the VTI sim IV. In: New development in driving simulation design and experiments: driving simulation conference Europe 2014 proceedings, pp 4.1–4.7
17. Korondi PZ, Gyeviki J (2006) Robust position control for a pneumatic cylinder. In: 12th International power electronics and motion control conference, pp 513–518. https://doi.org/10.1109/EPEPEMC.2006.283212
18. Lee J (2000) Investigation of quality indices of in-parallel platform manipulators and development of web based analysis tools. PhD Dissertation, University of Florida, Gainesville, FL
19. Lee J, Duffy J, Hunt K (1998) A practical quality index based on the octahedral manipulator. Int J Robot Res 17(10):1081–1090
20. Liu XJ, Wang J, Gao F, Wang LP (2002) Mechanism design of a simplified 6DOF 6-RUS parallel manipulator. Robotica 20:81–91. https://doi.org/10.1017/S0263574701003654
21. Liu Y, Wu H, Yang Y, Zou S, Zhang X, Wang Y (2018) Symmetrical workspace of 6-UPS parallel robot using tilt and torsion angles. Math Probl Eng. https://doi.org/10.1155/2018/6412030
22. Merlet JP, Gosselin C (2008) Parallel mechanisms and robots. Springer Handbook of Robotics, pp 269–285. https://doi.org/10.1007/978-3-540-30301-5_13
23. Mihailidis A, Samaras Z, Nerantzis I, Fontaras G, Karaoglanidis G (2009) The design of a formula student race car: a case study. Proc Instit Mech Eng Part D J Automob Eng 223:805–818. https://doi.org/10.1243/09544070JAUTO1080
24. Milica L, Andrei G (2017) Particularities of fully-parallel manipulators in 6-DOFS robots design: a review of critical aspects. In: MATEC web of conferences, vol 94, p 05003. https://doi.org/10.1051/matecconf/20179405003
25. Pierrot F, Uchiyama M, Dauchez P, Fournier A (1990) A new design of a 6-DOF parallel robot. JRM 2:308–315
26. Pradipta J, Klünder M, Weickgenannt M, Sawodny O (2013) Development of a pneumatically driven flight simulator Stewart platform using motion and force control. In: 2013 IEEE/ASME international conference on advanced intelligent mechatronics, pp 158–163
27. Raghavan M (1993) The Stewart platform of general geometry has 40 configurations. ASME J Mech Des 115(2):277–282. https://doi.org/10.1115/1.2919188
28. Roberts KM (1980) The FHWA highway driving simulator. Public Roads 44:97–102
29. Siddharth A, Swapnesh A, Shivankar, Udai AD (2019) An IoT based 6-DOF parallel robot for industrial application. In: Digital transformation through Industry 4.0, AITISM, 1, vol 1, p 500. 978-93-5361-329-7
30. Siddharth A, Udai AD (2020) WebGL Stewart simulation. http://ashishsiddharth.gitlab.io/stewart. Accessed Feb 2020
31. Stewart D (1965) A platform with six degrees of freedom. Proc Instit Mech Eng 180(1):371–386
32. Tsai L (1999) Robot analysis, the mechanics of serial and parallel manipulators. John Wiley and Sons, New York, U.S.A
33. Udai AD (2020) Racing car simulator using servo-pneumatic actuated gough-Stewart platform. https://youtu.be/keE5ChhCodA. Accessed 15 June 2023
34. Udai AD, Saha SK, Dayal A (2018) Overlaid orthogonal force oscillations for robot assisted localization and assembly. ISME J Mech Des 2(1):9–25
35. Valdiero AC, Ritter CS, Rios CF, Rafikov M (2011) Nonlinear mathematical modeling in pneumatic servo position applications. Math Probl Eng 2011:1–16. https://doi.org/10.1155/2011/472903
36. Wakasawa Y, Ito Y, Yanada H (2014) Dynamic behaviors of pneumatic cylinder (friction and vibration characteristics). Int J Fluid Power 8:60–65. https://doi.org/10.5739/JFPSIJ.8.60

37. Wang J, Pu JS, Moore PR (1999) A practical control strategy for servo-pneumatic actuator systems. Control Eng Pract 7:1483–1488. https://doi.org/10.1016/S0967-0661(99)00115-X
38. Zhang Y, Sun Y, Chen Y (2014) A framework for ergonomics design of transport category airplane cockpit. Proced Eng 80. https://doi.org/10.1016/j.proeng.2014.09.113
39. Ziegler IG, Nichols NB (1942) Optimum setting for automatic controllers. Trans ASME 64:759–768

Design of Flexure Robotic Hand for Teleoperation

Mohammad Zubair, Shubham Bhandari, and Indranath Chatterjee

Abstract Robots are getting deployed at home and in industries, becoming an integral part of our daily lives. Sophisticated mechanisms are required to make the joints of the robots compact and simple. In this paper, we have developed a flexure joint for the fingers of the robotic hand. The flexure joint is arranged in series to make a three-DOF underactuated finger. Five fingers are arranged to make a robotic hand. Castigliano's theorem is used to develop the deformation related to the applied loads, using the equation the flexure mechanism was designed. Pseudo-rigid model is developed to determine the stiffness of the flexure mechanism. A virtual model of the flexure mechanism is developed in a CAD environment. Finite Element (FE) simulation is conducted to the different loading conditions to evaluate the performance of the flexure mechanism. It was seen that the flexure mechanism is flexible about the rotational axis and stiff about the other axes. The stiffness model developed using the pseudo-rigid model has a good correlation with the FE simulation model.

Keywords Robotic hand · Flexure joint · Curved flexure · Castigliano's theorem

1 Introduction

It has been a while since robots are deployed for doing repetitive tasks. The performance of robots has increased over time. For unstructured work where the environment is dynamic, the robot can be deployed with unintuitive controls like remote joysticks [1]. Similarly, grasping and dexterous manipulation can be achieved by

M. Zubair (✉)
Department of Mechanical Engineering, Adamas University, Kolkata, India
e-mail: mdzubair87@gmail.com

S. Bhandari
Software Development Engineer-I, HashedIn by Deloitte, Bengaluru, India

I. Chatterjee
Department of Computer Science, Manchester Metropolitan University, Manchester, United Kingdom

S. K. Saha and K. Rama Krishna (eds.), *Recent Developments in Multibody Dynamics*, Lecture Notes in Mechanical Engineering,
https://doi.org/10.1007/978-981-97-7525-5_5

using only two actuator systems [2], where the mechanism supports performing tasks more realistic like humans. The manipulator can perform actions as per the user's hand gestures with minimum delay. This can be deployed at the point of concern with being controlled from a remote location. The robotic hand can be used in activities like bomb disposal or dealing with radioactive substances [3]. Unlike programmed robots, the robotic hand can leverage human intuition.

To develop a robotic hand having a minimal joint tolerance flexure mechanism is a possible choice where the bearings at the joints can be removed. A flexure mechanism has may advantage over a conventional mechanism, where friction losses, wear, and tear are absent [4]. Flexure mechanisms are deployed for MEMS, nano-manipulators, and medical devices [5]. Even in the case of a robotic system flexure mechanism can be deployed to develop a mechanism [6–9].

In this paper, we have developed a robotic hand having flexure mechanism, which is a gradual development from the conventional model developed. The finger in the robotic hand has three curved flexures connected in series and is tendon-driven by the actuator member. The flexures present in the hand ensure that the finger is retraced back to its original configuration once the load is removed. The design of the flexure is discussed in Sect. 2. FE simulation is conducted to see the performance of the flexure system and is discussed in Sect. 3. Results and discussions are subsequently covered.

2 Design of Robotic Hand

Our first version of the robotic hand has five fingers that are under-actuated using the tendon-driven system as shown in Fig. 1a. Each finger has three joints that are formed by spring-loaded pin joints. They are internally connected by the wire (as a tendon) and are driven by an actuator. The setup was fabricated in parts using the 3D printer and was assembled. The setup was used to demonstrate the teleoperation using the hand gesture [10].

Sooner, some of the design challenges were identified where the teleoperated hand motion was not accurately achieved. This was majorly due to poor joint design, the joint was having clearance for placing bearings, springs, and pins to form the joints. The three joints in series formed in a finger have a noticeable amount of unactuated motion (undesirable). This leads to the redesign of the robotic hand using flexure.

The robotic hand was designed using flexure joints, which is a substitute for conventional revolute joints present in the first version. The flexure was designed such that it behaves as an elastic rotational joint and the three segments of the finger can be one entity during fabrication. The curved flexure is used to form the joints in the fingers, which provided rotational motion and elastic deformation. Thus, the dependency on bearings and springs to form the revolute joint is taken care of.

For finger design, three curved flexures are used and are connected in series to form the joints of the fingers. The joints in the fingers are: the distant flange is connected to the middle flange, the middle flange is connected to the proximal flange, and

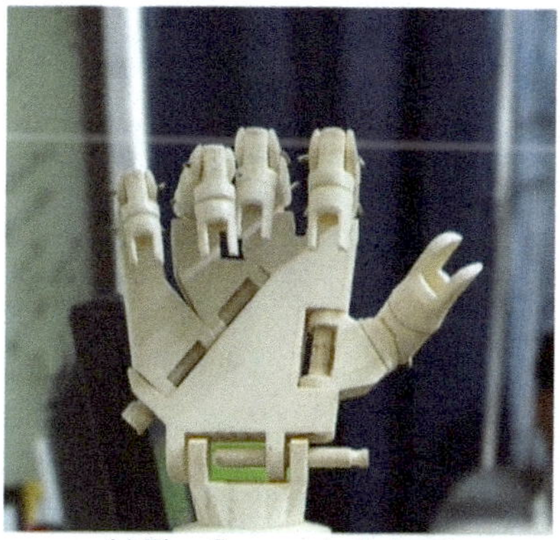

(a) Five-fingers robotic hand

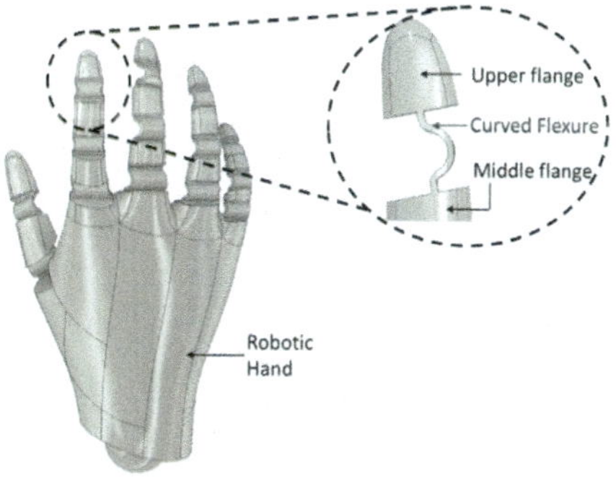

(b) Flexure joint-based robotic hand

Fig. 1 Development of robotic hand for teleoperation

the proximal flange is connected to the metacarpal, and is connected by the curved flexure. Five such fingers are used to form the robotic hand as shown in Fig. 1b. This version of the robotic hand has the ease to manufacture using 3D printer technology as no assembly is required.

2.1 *Design of Flexure*

A curved flexure that connects the two rigid bodies is shown in Fig. 2a. The flexure can be designed using Castigliano's theory. The forces and moments acted on one end can be considered as acting along *X*, *Y*, and *Z*-axis as shown in Fig. 2c. Based on these applied forces and moments, the deflection on the flexure can be expressed as:

$$\mathrm{U} = \int \frac{M_y^2}{2EI_y} ds + \int \frac{M_z^2}{2EI_z} ds + \int \frac{\alpha F_y^2}{2GA} ds + \int \frac{\alpha F_z^2}{2GA} ds + \int \frac{F_x^2}{2EA} ds + \int \frac{M_x^2}{2GJ} ds \tag{1}$$

where

M_x, M_y, and M_z are the moment acting along *X*, *Y*, and *Z*-axis.
F_x, F_y, and F_z are the forces acting along *X*, *Y*, and *Z*-axis.
I_x, I_y and the moment of inertia about *X* and *Y*-axis.
E is the modulus of elasticity.
G is the modulus of rigidity.
A is the area of cross-section.
J is the polar moment of inertia.

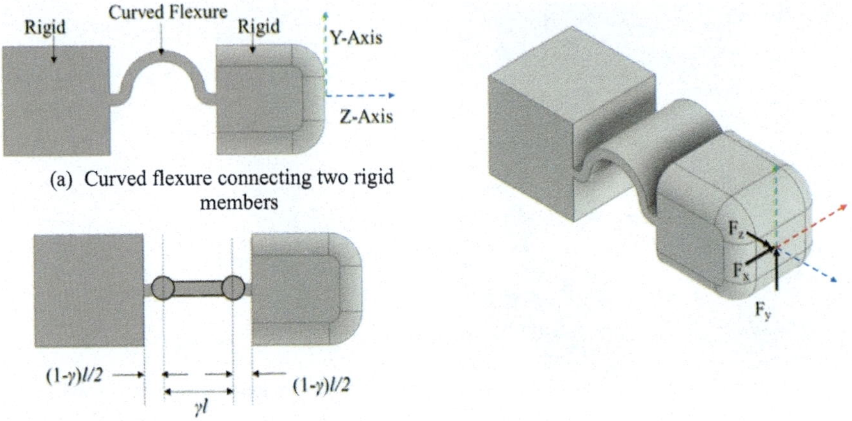

(a) Curved flexure connecting two rigid members

(b) Pseudo-rigid model of the curved flexure

(c) Curved flexure connecting the two rigid bodies subjected to forces and moments

Fig. 2 The curved flexure model for the design of joint

The angular and linear deflection can be formulated accordingly for the flexure mechanism. Using the known parameters for the system, the flexure has been designed. Where the thickness of the flexure for the angular rotation (as hinge) was determined as 0.5 mm, the curved length of 15 mm, and breadth of 5 mm.

Based on the design model, the stiffness of the flexure can be estimated using the pseudo-rigid theory. The curved flexure can be converted to equivalent pseudo-rigid model, consisting of three rigid links connected by torsional springs as shown in Fig. 2b. The torsional stiffness can be expressed as

$$K_z = \rho K_\theta \frac{EI_{zz}}{L} \tag{2}$$

where

K_θ is a constant and determined as 2.65 [11].
ρ is a constant and determined as 0.85 [11].
E is modulus of elasticity.
I_{zz} is moment of inertia about Z-axis.
L is the length of the flexure.

Based on the pseudo-rigid model, the stiffness of the flexure was determined for the known values of the system as 1.7 Nmm/degree.

3 FE Simulation

For determining the performance of the flexure, Finite Element (FE) simulation was conducted. The mechanism was modeled in Autodesk Inventor based on the dimension estimated using Castigliano's theorem. The simulation was further conducted using the FE simulation package of the Autodesk Inventor module. The meshing of the model was done using tetrahedron elements. The boundary condition was applied where one end was fixed to the base and the other end was subjected to the forces and moments. A force and a moment were applied at one instance and the deflection was measured. Thus, in total, six simulations were conducted for the load subjected along one of the axes at a time. Figure 3 shows the deformation contour plot of the flexure system, when force is applied along the X, Y, and Z-axis similarly, for the moments applied along the X, Y, and Z-axis.

Based on the simulation of the flexure mechanism, a flexure hand was developed in the CAD environment, as shown in Fig. 4a. The performance of the flexure hand was conducted using the FE package. The wrist was grounded, and loads were applied to all the fingertips. Based on the applied load, the deflection of the fingers was evaluated as shown in Fig. 4b.

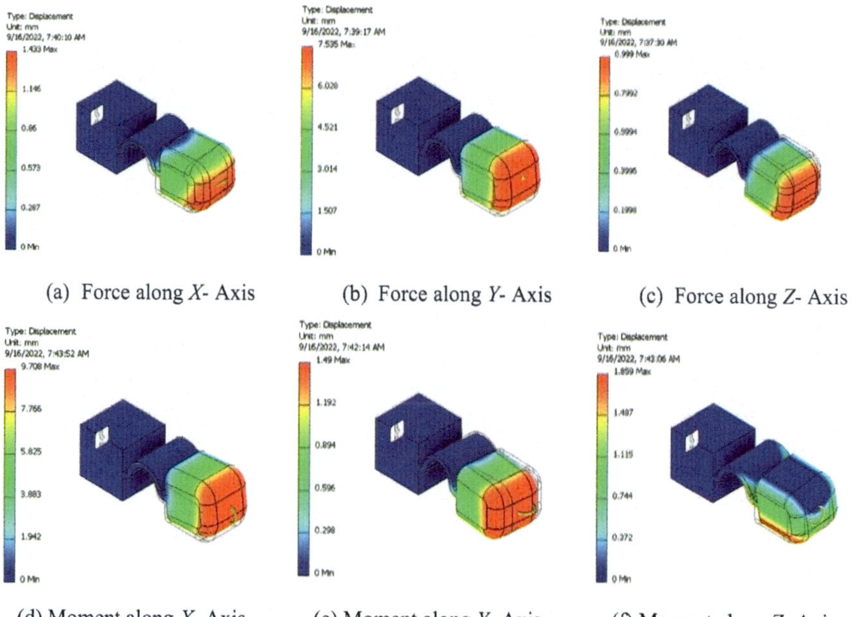

(a) Force along X- Axis (b) Force along Y- Axis (c) Force along Z- Axis

(d) Moment along X- Axis (e) Moment along Y- Axis (f) Moment along Z- Axis

Fig. 3 Contour plot of the flexure system subjected to forces and moments

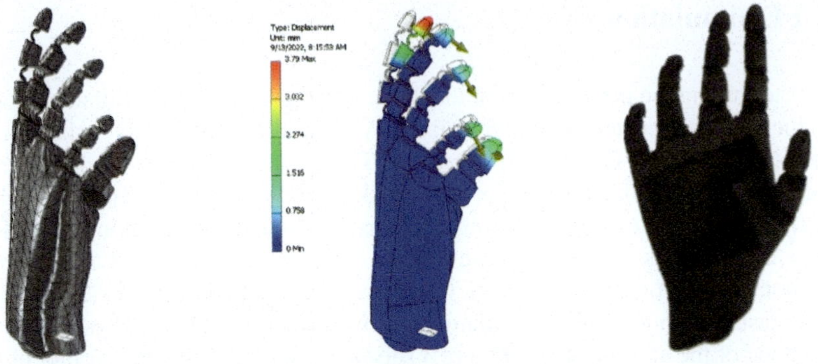

(a) Mesh, model of the flexure robotic hand.

(b) Deflection contour plot when loads were applied on the fingertip

(c) Developed model using a 3D printer

Fig. 4 The flexure robotic hand model developed in a CAD environment and the fabricated model using a 3D printer

4 Result

The FE simulation deflection contour plot data were used to determine the stiffness of the flexure system when it is loaded along the *X*, *Y*, and *Z*-axis. The linear stiffness along *X*-axis is 3.4 N/mm, along *Y*-axis is 0.664 N/mm, and along *Z*-axis is 5.05 N/mm. The torsional stiffness was evaluated along the *X*-axis as 1.79 Nmm/degree, along *Y*-axis as 11.70 Nmm/degree, and along *Z*-axis as 9.38 Nmm/degree. It can be seen the flexure mechanism is flexible about *Y*-axis for the linear deformation case.

The flexibility of the *Y*-axis compared to *X*-axis is nearly five times and the flexibility of the *Y*-axis compared to *Z*-axis is nearly eight times. It can be seen the flexure mechanism is flexible about *X*-axis for the angular deformation case. The flexibility of the *X*-axis compared to the *Y*-axis is nearly seven times and the flexibility of the *X*-axis compared to *Z*-axis is nearly five times (Fig. 5).

The FE simulation result and the pseudo-rigid model result were found to have a good correlation. Where the torsional stiffness about *X*-axis was evaluated as 1.7 Nmm/degree using pseudo-rigid model and 1.8 Nmm/degree from FE simulation. Based on this study, the flexure robotic hand is 3D printed as shown in Fig. 4c.

5 Discussion

A flexure mechanism is designed for the robotic hand. The flexure mechanism has initial curvature as identified for the system. The design of the flexure was done using Castigliano's theorem. A pseudo-rigid model is developed to determine the stiffness of the mechanism. FE simulation is conducted for the applied load for six different

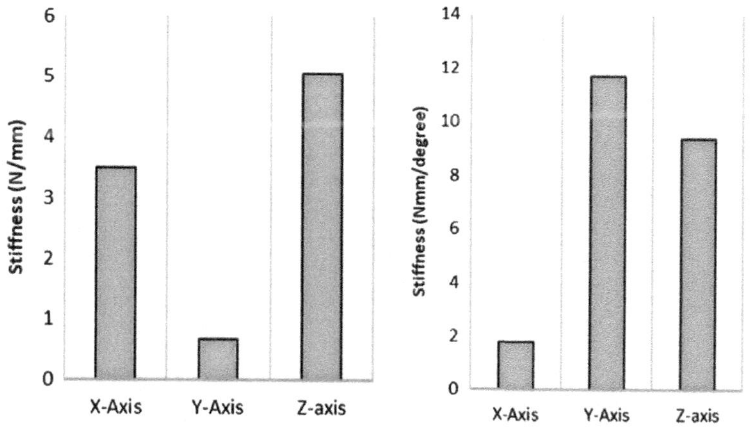

(a) Linear stiffness plot (b) Torsional stiffness plot

Fig. 5 The linear and torsional stiffness of the flexure system

cases. Based on the FE simulation, the result shows that the flexure mechanism is flexible about one of the axes and stiff about the other axes. The pseudo-rigid model and the FE result were found to have a good correlation. The flexure robotic hand is developed using a 3D printer.

Based on the FE results, the tendon drives actuation system can be designed. The performance of the robotic hand can be tested for the teleoperation cases, which will be carried out gradually.

References

1. Meattini R et al (2018) An sEMG-based human–robot interface for robotic hands using machine learning and synergies. IEEE Trans Compon Packag Manuf Technol 8(7):1149–1158
2. Sun B-Y et al (2021) Design principle of a dual-actuated robotic hand with anthropomorphic self-adaptive grasping and dexterous manipulation abilities. IEEE Trans Robot
3. Khan S et al (2017) Gesture controlled prosthetic arm. Int J Adv Res Dev 2(6):56–59
4. Jones RV (1962) Some uses of elasticity in instrument design. J Sci Instrum 39(5):193
5. Smith ST (2000) Flexures: elements of elastic mechanisms. CRC Press
6. Slocum AH (1992) Precision machine design. Society of Manufacturing Engineers
7. Awtar S, Ustick J, Shiladitya S (2013) An XYZ parallel-kinematic flexure mechanism with geometrically decoupled degrees of freedom. J Mech Robot 5(1)
8. Fischer FL (1981) Symmetrical 3 DOF compliance structure. U.S. Patent No. 4,447,048
9. Bednorz JG et al (1985) Piezoelectric XY-positioner. U.S. Patent No. 4,520,570, 4 Jun 1985
10. Bhandari S et al (2021) Design of robotic hand and wireless glove to tele-operate. In: Advances in robotics-5th international conference of the robotics society
11. Howell LL (2013) Compliant mechanisms. In: 21st century kinematics, vol 710. Springer, pp 189–216

Design and Development of a Novel Rotary Actuator Based on Shape Memory Alloy and Permanent Magnet System

Deep Singh, Rutupurna Choudhury, Yogesh Singh, and Santhakumar Mohan

Abstract This paper presents a novel rotary actuator's design concept based on shape memory alloy (SMA) springs and permanent magnet systems. The proposed actuator is so designed that it can operate even in a vacuum. The presence of permanent magnets leads to efficient and quick rotation motion without any direct contact between the actuation system and passive disc rotation. The presence of a smart actuator (SMA) system helps build a lightweight actuator. The presented actuator converts the translational motion of the SMA springs in the bi-directional rotation of the passive disc. The contraction of the SMA could directly control the rate of rotation.

Keywords Shape memory alloy (SMA) · Permanent magnet · Rotary actuator · Nitinol (NiTi)

1 Introduction

Several researchers have already developed a large number of linear to rotary actuators that incorporate several simple and complex techniques. The application of linear and rotary actuators is infinite. The demand for micro- and nano-level actuation tracking has expanded due to recent micro and nanoscale production developments. Industry 4.0 has also contributed to automation and continuous in-process

D. Singh · R. Choudhury · Y. Singh (✉)
Department of Mechanical Engineering, National Institute of Technology Silchar, Assam 788010, India
e-mail: yogesh@mech.nits.ac.in

S. Mohan
Department of Mechanical Engineering, Indian Institute of Technology Palakkad, Kerala 678557, India
e-mail: santhakumar@iitpkd.ac.in

S. K. Saha and K. Rama Krishna (eds.), *Recent Developments in Multibody Dynamics*, Lecture Notes in Mechanical Engineering,
https://doi.org/10.1007/978-981-97-7525-5_6

monitoring of various parameters, including the translational and rotational movement in multiple domains, including manufacturing. Linear to rotary actuation is a very significant parameter for industrial applications.

Jin et al. [1] presented a novel 2 DOF linear and rotary Halbach permanent magnet actuator. Krebs et al. [2] modeled and analyzed an electromagnetic actuator involving linear and rotary motions. Ni et al. [3] proposed a novel partitioned-stator linear-rotary permanent magnet actuator (PS-LRPMA) having excellent electromagnetic performance, such as higher thrust and torque density. Guo et al. [4] also developed a novel linear-rotary permanent magnet actuator with an independent magnetic circuit structure. Xu et al. [5] investigated a double stator linear-rotary permanent magnet motor (DSLRPMM) with permanent magnets arranged orthogonally. Viscuso et al. [6] also developed a portable rotary actuator using a magnetic shape memory alloy, which is actuated by Joule's effect. Multiple analysis proved its application in various fields, including biomedical, robotic, aerospace, automotive. Kalimullina et al. [7] analyzed a micro-actuator developed based on a composite bilayer of rapidly quenched NiMnGa-based ferromagnetic SMA and an elastic Platinum layer. Hubert et al. [8] presented magnetic SMAs as an excellent candidate for their application in the mechatronics domain. It can be used effectively for the development of micro-positioning stages. Singh et al. [9] and Choudhury et al. [10] have also implemented the SMA (Nitinol) spring for linear actuation in a planar parallel micro-motion positioning stage.

SMAs are smart materials that attain their memorized shape in the presence of external parameters such as stress, temperature, and electric and magnetic fields. SMAs, being numerous good characteristics, are widely used as actuators in multiple domains, including biomedical, commercial, aerospace industries, etc. SMAs are compact with higher power per unit mass, undergo silent operation [9, 11–13]. Nitinol (NiTi) is one of the SMA having a composition of 50% Nickel and 50% Titanium. Nitinol (at detwinned state) undergoes shape change under thermal loading due to crystal reorientation from the monoclinic crystal structure (martensite region) to body-centered cubic crystal structure (austenite region). It is biocompatible and also activates at lower voltages [9, 11, 12, 14].

Due to its outstanding power-to-weight ratio, shape memory alloys are useful as actuators in applications where weight reduction is a priority. Another factor driving current research is the growing need for passive rotational actuators in various applications, where active actuators cannot be used because of the working environment. Therefore, there is a strong need for research and development to produce SMA-driven active rotary actuators that allow passive rotary actuators to rotate with the help of a carefully arranged array of permanent magnets, enabling operation in a variety of environmental conditions. The present study proposes a novel smart rotary actuator integrated with permanent magnets and Nitinol SMA springs and evaluates the bi-directional angular rotation experimentally with the actuation of SMA springs. This paper is further arranged in the following order: (2) Concept design of the proposed novel actuator system; (3) Performance of Nitinol SMA actuator; (4) Experimental setup of novel rotary actuator; (5) Results and discussion; and (6) Conclusion.

2 Concept Design of the Proposed Novel Actuator System

The proposed permanent magnet-based novel rotary actuator, as depicted in Fig. 1, consists of an active and a passive cylindrical disc positioned at a fixed height from the base and separated at a certain distance. Two NiTi SMA springs connect the active disc to the base with the help of a thread, as depicted in Fig. 1, for bidirectional rotation of the active disc. The SMA spring contracts as DC power is supplied across it. The rotation of the active disc is possible with the power supply. Furthermore, the disc's rotation direction would be towards the active SMA spring (that is, whichever direction spring energizes, the disc would rotate in that side). This action is due to the contraction of the active SMA spring. Simultaneously, the other SMA spring elongates due to the force exerted by the actuation of the active spring. Since permanent magnets are attached to both discs, as depicted in Fig. 1, the passive disc rotates by rotation of the active disc due to spring actuation. It is due to the presence of magnetic force between the magnets.

The proposed actuator has been designed to have no physical contact between the active and passive cylindrical discs. The actuator operates even when both discs are positioned in different environments. The separation between the discs (or magnets) is a significant parameter. The force of attraction between two permanent magnets is inversely proportional to the square of the distance between them. The force exerted by two magnetic poles is given by Eq. (1).

$$F = \frac{\mu q_{m1} q_{m2}}{4\pi r^2} \tag{1}$$

where F (Newton) is the force between two magnetic poles of magnitudes q_{m1}, q_{m2} (ampere-meter) separated by a distance r (meter), μ (Tesla meter per ampere) is the permeability of the intervening medium.

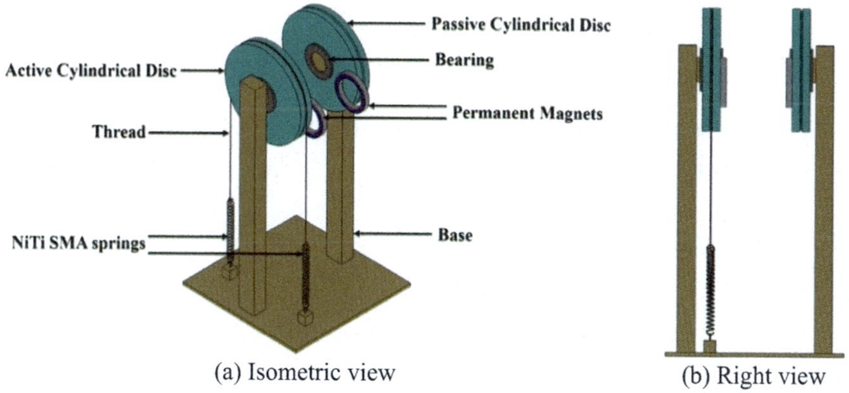

Fig. 1 Proposed permanent magnet-based SMA-operated actuator

3 Performance of SMA Actuator

SMA-based actuators work on the principle of the joule-heating method. The temperature of the actuator rises when direct current (DC) is applied across the SMA due to the electrical resistance of the SMA-based springs. Because of the temperature rise, the pre-stretched SMA actuator undergoes a phase transformation from a lower-symmetry martensitic phase to a higher-symmetry austenite phase, resulting in the contraction of the SMA spring.

An Arduino Uno microcontroller, a Baumer Ultrasonic sensor, a programmable power supply, and a thermocouple constitute the experimental setup. The SMA spring is loaded with the input voltage that the controller has calculated and that has been supplied by a programmable power source. The linear displacement of the SMA spring is measured by the Baumer ultrasonic non-contact displacement sensor. To gauge the SMA spring's temperature, a thermocouple is fastened. Figure 2 depicts the complete experimental setup for the SMA actuator. The experiment includes the fixed base and rail guide with the help of a fixed block at the two ends of the base. A slider block is installed at the rail guide which connects the SMA springs. The SMA spring is connected to the slider block on one end and the fixed block on the other. The SMA spring's temperature is gauged using a thermocouple. The MATLAB Simulink software and Arduino microcontroller enable the Baumer displacement sensor to measure the SMA spring's deflection.

The experimental results of the SMA actuator are presented in Figs. 3 and 4. The experiment is conducted with a maximum voltage range of 0–12 V and a current of 0–3.7A. The time histories of the displacement are shown in Fig. 3. It is shown that with the increase in the current value, the spring displacement is gradually increased and decreased with the decrease in the current value. The maximum displacement of the SMA actuator is attained up to 105 mm (the initial length of the SMA spring is 140 mm) at the current value of 3.7A. Again, it is going to the original position with the decrease in the current value. Both the heating and cooling curves are presented in Fig. 3. The force–time histories of the SMAS actuator are depicted in Fig. 4.

4 Experimental Setup of SMA-Actuated Novel Rotary Actuator

The experimental setup designed for the proposed novel SMA-actuated permanent magnet-based rotary actuator has been depicted in Fig. 5. Various devices and units integrated to develop the experimental setup have also been presented in Table 1.

The SMA springs considered for the experiment are Nitinol and are readily available. The spring consists of 19 helical windings of 0.75 mm wire diameter that could contract and elongate between 29 and 140 mm.

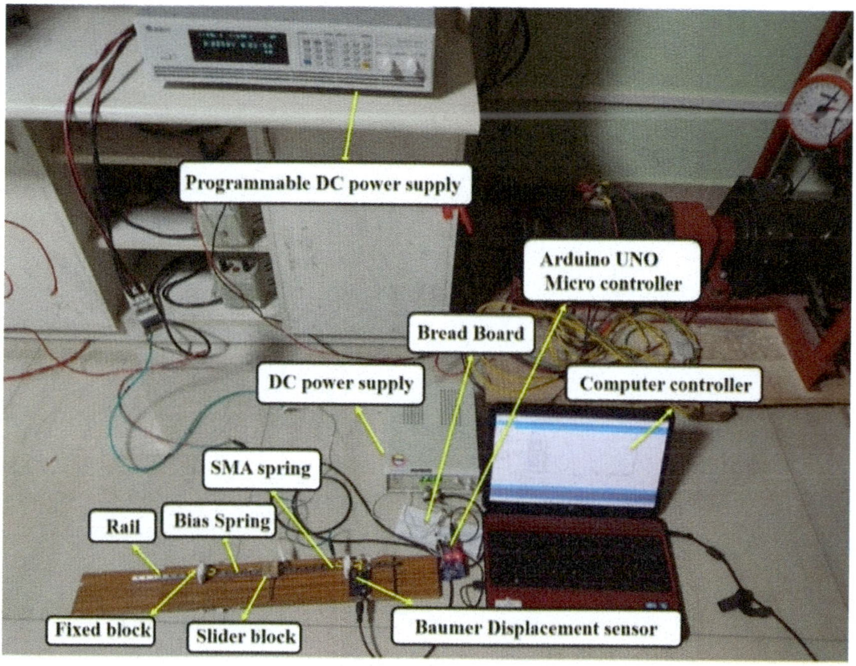

Fig. 2 Experimental setup of the SMA actuator

Fig. 3 Displacement–time histories of linear actuation-based SMA spring

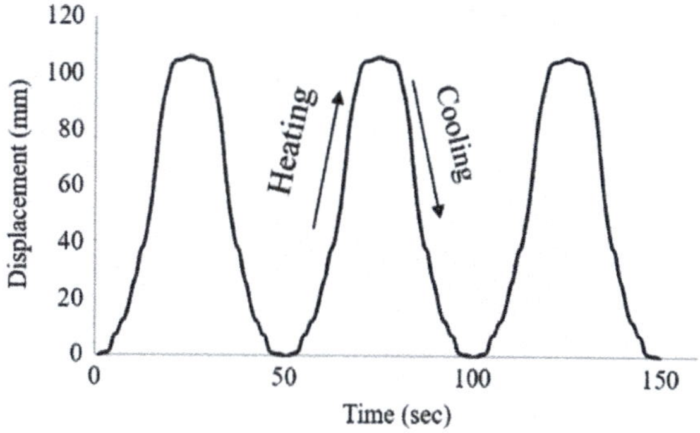

5 Results and Discussion

The SMA-based actuators are incredibly lightweight and undergo simple actuation. To conduct the experiment, the Joule heating process acts on SMA springs. As direct current is supplied through the SMA spring, the spring heats up due to its resistance. As temperature rises, SMA undergoes phase transition from the martensitic state to the austenitic state due to variations in crystal structure. The actuation of the SMA spring on the left, as illustrated in Fig. 5, rotates the active cylindrical disc

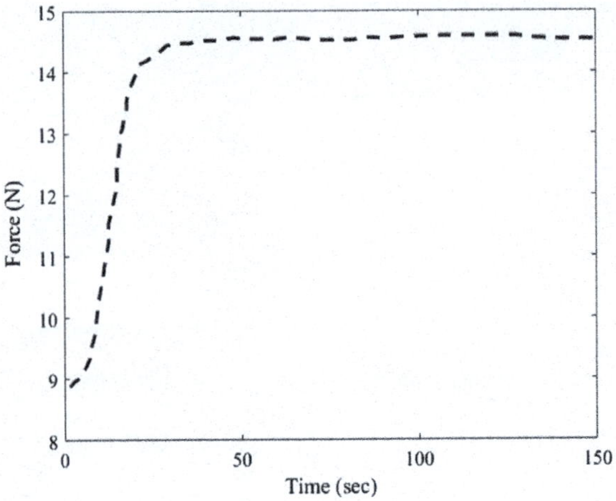

Fig. 4 Force–time histories of the linear actuation-based SMA spring

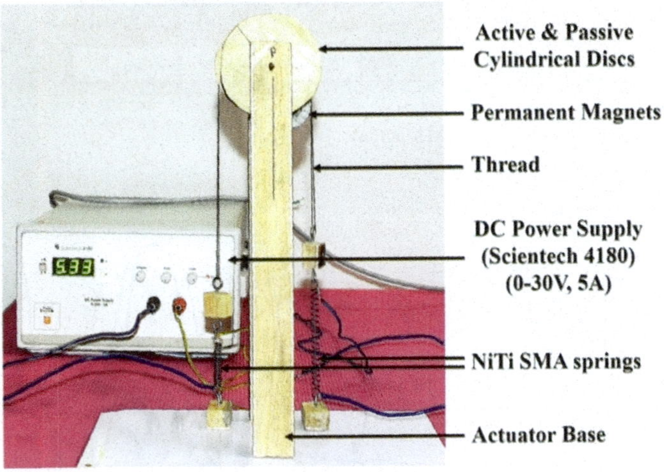

Fig. 5 Experimental setup of the proposed novel actuator

counterclockwise. Similarly, the actuation of the SMA spring on the right rotates the active cylindrical clockwise. Due to magnetic force, the passive cylindrical disc rotates in the same direction as the active cylindrical disc.

Five sets of experiments were conducted to rotate the proposed actuator bidirectionally. The experimental setup of the developed novel actuator is presented in Fig. 5. Direct current was supplied to one of the SMA spring and the angular rotation of the cylindrical discs was measured. Similarly, the current was supplied to the other SMA spring, and the opposite direction's angular rotation was measured. The plot of the experimental result is presented in Fig. 6. The experiment results mean angular

Table 1 List of devices used for the design of experimental setup

Serial number	Device	Specification	Quantity
1	Proposed rotary actuator model consisting of a fixed base and two cylindrical discs	A fixed base and two cylindrical discs (made from wood)	1
2	Variable DC power supply	Scientech 4180 (Output: 0–30 V, 0–5 A)	1
3	Permanent magnet	–	2
4	String (or thread)	–	1
5	Nitinol SMA spring	19 helical windings	2

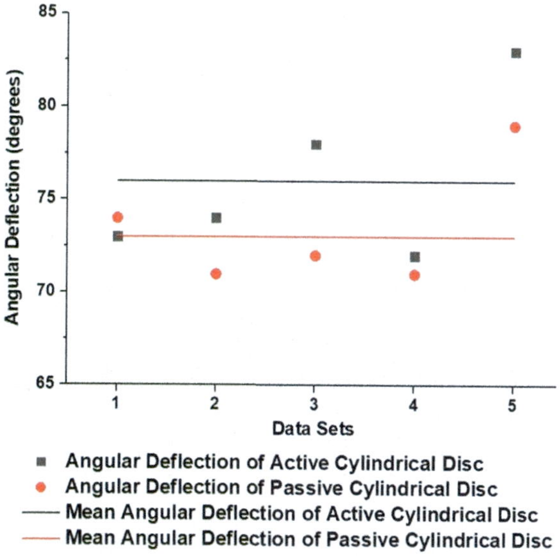

Fig. 6 Angular deflection of the active and passive cylindrical disc

rotation of 76° and 73.4° for the active and passive discs with a standard deviation of 4.52° and 3.36°, respectively.

The precision, accuracy, and sensitivity characteristics of the proposed actuator primarily depend on the following parameters:

- Distance between the magnets
- Magnitudes of the magnetic poles
- The radius of the cylindrical disc
- Contraction rate of SMA spring.

Proper selection of magnets can overcome the issue of the first two parameters. The remaining two parameters are based on the design considerations. The proposed actuator's desired characteristics could be achieved by varying the radius of the cylindrical discs and the SMA contraction rate. The angle of rotation (θ) of the cylindrical discs (of radius R) achieved due to the length of contraction (L) of the active SMA spring can be represented by Eq. (2).

$$\theta = \frac{360 \times L}{2\pi R}(\text{in degrees}) \tag{2}$$

As the angle of rotation is inversely proportional to the cylindrical disc's radius, it can be stated that better sensitivity of the actuator could be achieved with an increase in the radius of the disc and vice-versa. Also, enhanced sensitivity could be accomplished with a reduced SMA spring contraction rate.

Since one magnet is attached to each disc, owing to the magnet's immense gravitational force, the discs are unable to achieve full rotation. Therefore, for improved efficiency, equal numbers of multiple smaller permanent magnets could be fixed near each disc's circumference.

6 Conclusion

The major conclusions of the present study of the proposed smart actuated novel magnetic rotary actuator have been outlined below:

- The angular rotation of the cylindrical discs is directly proportional to the contraction length of the active SMA spring.
- For a set of selected magnets, the precision, accuracy, and sensitivity of the proposed smart magnetic rotary actuator depend indirectly on the radius of the cylindrical discs and directly on the contraction rate of the SMA spring.
- The magnitude of the magnetic poles and the magnets' distance also plays a vital role in sensitivity.
- Instead of bigger magnets, implementing multiple smaller permanent magnets along the periphery or circumference of the cylindrical discs would boost the proposed actuator's performance.
- Accurate rotation with micron-level precision is possible with the help of a SMA spring-based actuator.
- The proposed rotary actuator could operate even when both the active and passive cylindrical discs are present in two separate mediums as long as the magnetic field of one magnet affects the other magnet.

Acknowledgements The presented work in the paper is supported and funded by the CSIR-EMR-II, India (Number 22(0831)/19/EMR-II) and Start-up Research Grant, SERB, DST, India (Number SRG/2020/000491).

References

1. Jin P, Fang S, Lin H, Wang X, Zhou S (2012) A novel linear and rotary Halbach permanent magnet actuator with two degrees-of-freedom. J Appl Phys 111(7):07E725
2. Krebs G, Tounzi A, Pauwels B, Willemot D, Piriou F (2008) Modeling of a linear and rotary permanent magnet actuator. IEEE Trans Magn 44(11):4357–4360
3. Ni H, Fang S, Yang H, Lin H (2016) A linear-rotary permanent magnet actuator with partitioned stator. In: 2016 IEEE vehicle power and propulsion conference (VPPC), Hangzhou, pp 1–6
4. Guo K, Fang S, Lin H, Huang Y, Zhang Y, Yang H (2016) A linear-rotary permanent magnet actuator with independent magnetic circuit structure. IEEE Trans Appl Supercond 26(7):1–6
5. Xu L, Lin M, Fu X, Li N (2016) Analysis of a double stator linear rotary permanent magnet motor with orthogonally arrayed permanent magnets. IEEE Trans Magn 52(7):1–4
6. Viscuso S, Pittaccio S (2012) Design and implementation of a portable amagnetic shape memory rotary actuator. J Intell Mater Syst Struct 24(4):454–472
7. Kalimullina E, Kamantsev A, Koledov V, Shavrov V, Nizhankovskii V, Irzhak A, Albertini F, Fabbrici S, Ranzieri P, Ari-Gur P (2014) Magnetic shape memory microactuator. Phys Status Solidi C 11(5–6):1023–1025
8. Hubert A, Calchand N, Le Gorrec Y, Gauthier J-Y (2012) Magnetic Shape Memory Alloys as smart materials for micro-positioning devices. Adv Electromagn 1(2):75–84
9. Singh D, Choudhury R, Singh Y, Mukherjee M (2020) Development and workspace analysis of smart actuation based planar parallel robotic motion stage. In: IOP conference series: materials science and engineering, vol 912, p 032063
10. Choudhury R, Singh D, Kumar A, Singh Y, Sahoo CK (2021) Development and workspace study of a 4-PP planar parallel XY positioning stage using SMA actuators. In: Pandey K, Misra R, Patowari P, Dixit U (eds) Recent advances in mechanical engineering. Lecture notes in mechanical engineering. Springer, Singapore, pp 21–31
11. Singh D, Singh Y, Mukherjee M (2021) Behaviour of NiTi based smart actuator for the development of planar parallel micro-motion stage. In: Kalamkar V, Monkova K (eds) Advances in mechanical engineering. lecture notes in mechanical engineering. Springer, Singapore, pp 221–228
12. Singh Y, Mohan S (2017) Development of a planar 3PRP planar parallel manipulator using shape memory alloy spring based actuators. In: Proceedings of the advances in robotics (Proceeding AIR17), vol 10, pp 1–6
13. Mikov L, Medveck-BeÅov S, Kelemen M, TrebuÅa F, Virgala I (2015) Application of shape memory alloy (SMA) as actuator. METABK 54(1):169–172
14. Hirose S, Ikuta K, Umetani Y (1988) Development of shape-memory alloy actuators. Performance assessment and introduction of a new composing approach. Adv Robot 3(1):3–16

UAV Landing on General Moving Platforms Without Markers

Sagar Dalai, Kanishk Vishwakarma, Kaushal Kishore, Dhruv Potdar, and Manan Arora

Abstract Multirotor Unmanned Aerial Vehicles (UAVs) and their ability to hover and maneuver in the air make them the best vehicle for applications requiring quick package deliveries. Our assumed example system is a collaboration of a UAV and a land vehicle capable of storing multiple packages and charging the UAV's batteries. Another example is quickly exploring some unknown region where the UAV scans the unexplored region and returns to the moving vehicle when required knowledge is gathered. An active research problem in such systems is about takeoff and landing the UAV on the moving vehicle. The landing problem, however, is quite challenging without unique markers like visual tags, hard-coded trajectory targets, etc. We present a markerless controller that uses only computer vision and calculates the optimum trajectory to land the UAV on the moving platform. This paper presents the work in a two-stage implementation process. First, the detection of a moving vehicle and finding the centroid by applying optical flow and contour technique on image feed from an RGB camera, and secondly, following the ground vehicle by matching the velocity by utilizing a visual servoing process followed by simple PID controller to descend in the smooth trajectory towards the landing point.

Keywords UAV · Optimal flow · Visual servoing

S. Dalai (✉) · K. Kishore
Integrated Circuits and Systems Group, CSIR-CEERI, Pilani 333031, Rajasthan, India
e-mail: mr.sagardalai@gmail.com

K. Kishore
e-mail: kaushal@ceeri.res.in

K. Vishwakarma · D. Potdar · M. Arora
Department of Electrical and Electronics, BITS Pilani, K.K Birla Goa Campus, Goa 403726, India
e-mail: f20190315@pilani.bits-pilani.ac.in

D. Potdar
e-mail: f20201489@pilani.bits-pilani.ac.in

M. Arora
e-mail: f20200363@pilani.bits-pilani.ac.in

S. K. Saha and K. Rama Krishna (eds.), *Recent Developments in Multibody Dynamics*, Lecture Notes in Mechanical Engineering,
https://doi.org/10.1007/978-981-97-7525-5_8

1 Introduction

Rapidly changing domains of UAVs with multiple vehicles, such as fixed-wing and multi-copter designs, improved sensor technology, and robust algorithms for navigation, control, and path planning, have enabled the use of UAVs in a variety of applications. The active application areas include monitoring, exploration, survey, and surveillance based on their ability and structure. However, short endurance or limited payload capacity remains the critical bottleneck in the effective utilization of UAVs. Precision landing of drones for independent battery changing/charging has been used as an alternate route to minimize downtime due to the low endurance of drones. Furthermore, a generalized method would be to track and land on a moving ground vehicle or vessel deck platform in extreme sea state conditions [1]. Therefore, much research has been done to develop a complete system for the precise and robust landing of UAVs on movable platforms. The application of drones in disaster management also requires package delivery, flood survivor tracking, etc., which often deals with dynamically moving targets. Tracking and landing a drone under these circumstances become essential for the rescue team [2].

With the increasing interest of delivery companies like Swiggy and Amazon towards autonomous robotic deliveries using UAVs, the UAVs inevitably have to land on moving platforms in some cases [3]. Instances such as charging batteries by land on a vehicle during a mission or when collecting more packages from the land vehicle without requiring to go back to the original workstation. Either way, landing a UAV on a moving platform is a tricky problem with unpredictable target trajectories, occlusions, controls without collision, and ambient noise such as wind or bad visuals due to smoke or rain. Currently, a lot of work has been done but using visual markers [4], which is impractical in real-life scenarios where the target landing site will often not have markers or anything that can be vaguely considered a marker. Because of the generalization of the problem, deep neural networks [5] may not be feasible to detect the target landing platform.

In this paper, a complete UAV system is presented, which is capable of autonomously landing on a moving target, mainly using an onboard computer and sensors for UAV localization. An RGB camera is used to identify the moving platform for landing and estimate the platform's location without a special tag or markers like AprilTag, ArUco marker, or custom markers.

2 Related Work

Significant advancements have been made in drone landing technology over the past decade, highlighting its importance in various drone applications. Typically, the landing target is identified using methods such as color segmentation, AprilTags [6], custom landing markers [7], ArUco markers [8], and color markers [9].

Additionally, deep learning techniques, including deep reinforcement learning, have proven effective in motion control and navigation, particularly with the successful application of the Deep Q- Network (DQN) technique in Atari games. To solve continuous state and action domain problems in the robotics field, Deep Deterministic Policy Gradients (DDPG) algorithm [10] played an important role in providing outstanding results. Similarly, Xie et al. [11] try to implement a model-free method based on a partially observable Markov decision process. Here, UAV learns the landing maneuver using the Deep Deterministic Policy Gradients (DDPG) algorithm and heuristic rule-based end-to-end neural network.

Two critical challenges in landing a UAV on a moving platform are as follows. (i) Pose estimation or localizing the landing platform and the UAV. (ii) Trajectory generation and navigation under the presence of disturbances like wind, light, and other uncertainties. The work done by Feng et al. [12] states that autonomous UAVs landing on the center of a mobile platform traveling at a speed of up to 12 m/s with an error of less than 37 cm under wind disturbances. Borowczyk et al. [13] made use of AprilTags, a visual fiducial dictionary-based system, and an Inertial Measurement Unit (IMU) and GPS receiver integrated into a moving target traveling at 14 m/s. However, there are a few limitations even in the fiducial marker attached to the landing platform should be within the Field of View (FoV) of the camera throughout the whole landing routine [14]. Researchers also have introduced sensor fusion techniques to include cameras, IMUs, and other sensors for a more precise and robust system.

3 Proposed Method

Multirotor UAV research done on landing the UAV on a moving platform focuses only on the control systems part, with the moving platform pose being presumably known either by using extrinsic fiducial markers or by intrinsically subscribing to the real-time landing platform's location data stream. We solve the vision part by detecting the landing site without any fiducial markers, making use of different methods like contour detection, optical flow, and object detection. We then use techniques like path following and visual servoing to move and multirotor and execute the landing behavior.

The overview of our approach consists of three stages. The initial task is locating and classifying the moving target and the landing platform. The computer vision stack uses optical flow to segment the moving objects and finds the center of the landing platform of the moving vehicle, which is explained in Sect. 3.1. In Sect. 3.2, trajectory generation, and getting up to the platform's speed and orientation-visual servoing is performed so that the UAV can track the vehicle and start following a safe target trajectory. Calculating a safe, collision-free touchdown maneuver. Once the target viewpoint is reached and the UAV is following the vehicle on the target trajectory, a landing maneuver is performed using a simple PID-based controller.

The steps are detailed below (we use computer vision and mathematics software libraries like OpenCV [15] and NumPy [16] for their optimized low-level computing methods).

3.1 Detection

1. HSV: It stands for Hue, Saturation, and value, where value represents the max values of RGB of the image. Converting the image to HSV will make it easier to isolate colors.

$$V = max\{R, G, B\} \tag{1}$$

$$S = \frac{V - min\{R, G, B\}}{V} \tag{2}$$

where, $S \in [0, 1]$

$$H_{deg} = \begin{cases} \frac{G-B}{V-min\{R,G,B\}} \times 60^\circ, & \text{if V} = \text{R \& G} \geq \text{B} \\ \left(\frac{B-R}{V-min\{R,G,B\}} + 2\right) \times 60^\circ, & \text{if V} = \text{G} \\ \left(\frac{R-G}{V-min\{R,G,B\}} + 4\right) \times 60^\circ, & \text{if B} = \text{V} \\ \left(\frac{R-B}{V-min\{R,G,B\}} + 5\right) \times 60^\circ, & \text{if V} = \text{R \& G} < \text{B} \end{cases} \tag{3}$$

2. Optical Flow: There are a few assumptions that need to be considered. Firstly, the pixel intensity of an object does not change between consecutive frames $I(x, y, t)$, and secondly, neighboring pixels have similar motion, i.e. (dx,dy) with respect to time should be the same.

$$I(x, y, t) = I(x + d_x, y + d_y, t + d_t) \tag{4}$$

By taking tailor series approximation on the right-hand side of the equation (4), we will get the Optical Flow equation, i.e. shown in equation (6).

$$\frac{\mathrm{d}I}{\mathrm{d}x}u + \frac{\mathrm{d}I}{\mathrm{d}y}v + \frac{\mathrm{d}I}{\mathrm{d}t} = 0 \tag{5}$$

$$f_x u + f_y v + ft = 0 \tag{6}$$

As per our second assumption, all the neighboring pixels will have similar motions. Using the Lucas-Kanade method will take a 3×3 patch, i.e. the nine points have the same motion. So we have (fx,fy, ft) for these 9 points and two unknown variables (u,v), which can be solved with the least square fit method. Below are the final two equation-two unknown problems.

$$\begin{bmatrix} u \\ v \end{bmatrix} = \begin{bmatrix} \sum_i f_{xi}^2 & \sum_i f_{xi} f_{yi} \\ \sum_i f_{xi} f_{yi} & \sum_i f_{yi}^2 \end{bmatrix}^{-1} \begin{bmatrix} -\sum_i f_{xi} f_{ti} \\ -\sum_i f_{yi} f_{ti} \end{bmatrix} \tag{7}$$

The above equations deal with small motions. So for large motion, we will use pyramids of motion, while going up in the pyramid, small motions will be replaced by large motion.
Uniformly distributed linearly normalized n optical flow vectors $x_i \in X$ in the image space having velocities $v_i \in V$ are used to estimate the centroid of the moving platform x_m in the image space by calculating the weighted average of the vectors with their velocities as shown in Eq. 8 given the standard deviation of velocities σ_V is above the manually tested threshold σ_o to toggle between landing mode and hovering mode.

$$x_m = \frac{\sum_{i=1}^{n} x_i v_i}{n} \tag{8}$$

3. Contour Detection:
Contour is a curve having the same color or intensity formed by joining the continuous points along the boundary. This will be used in recognition and to find the center of the vehicle. To achieve that, the HSV technique is used. Firstly, we convert the RGB image to HSV color space, then apply lower and upper threshold limits of H (Hue), S (Saturation), and V (RGB values) for a range of black colors, which will separate the target object from the background by masking the HSV image to get only black colors. We have converted the images into binary format for efficient contour detection. Inverting and applying the contour to the masked image will get the vehicle contour all (x,y) points. Below are the steps to get the contour of the vehicle.

   (a) Gradient-Base Edge Detection:
   This method detects the edges based on the image intensity profile. Basically, the edges are located on the maximum and minimum values of the first derivative result. So we will get the below Eq. (9) from the first partial derivative of the image function $I(u, v)$ along u and v-axes.

$$\nabla I(u, v) = \begin{bmatrix} \frac{\partial I}{\partial u}(u, v) \\ \frac{\partial I}{\partial v}(u, v) \end{bmatrix} \tag{9}$$

   (b) Edge Sharpening(Unsharp Masking): There are two steps to sharpen the edges. Firstly, mask 'M' is generated by subtracting the smoothed version of image I with kernel H (smoothing filter). Then increase the mask 'M' to the factor 'a' and add to the original image function I.

$$M \leftarrow H - (I \times \tilde{H}) = (I - \tilde{I}) \tag{10}$$

$$\tilde{I} \leftarrow (I + a.M) \tag{11}$$

3.2 Visual Servoing

1. Aligning to platform's centroid: If $\sigma_V < \sigma_o$, the UAV slowly moves upwards in a spiral until $\sigma_V > \sigma_o$, indicating that a potential landing site came into view (assuming it has a flat surface). Whenever $\sigma_V > \sigma_o$, pixel-wise distance between image's center x_c and platform's centroid x_m is calculated and the velocity v_{xy} in the $x - y$ plane is given along the log function of the calculated distance to result in minimum jerk proportional controller to minimize the distance between x_m and x_c.

$$v_{xy} = \frac{x_m - x_c + \varepsilon}{|x_m - x_c| + \varepsilon} \log(|x_m - x_c| + 1) \tag{12}$$

2. Landing maneuver: The closer the platform's centroid to the image's center, more the landing velocity, a behavior which is achieved by vertical velocity

$$v_z = -\frac{1}{\log(|x_m - x_c| + 1)} \tag{13}$$

   Due to non-ideal alignment maneuvering and fast landing behavior, it is difficult to have the platform always in view. To tackle this we assume the platform goes out of sight for small time periods not greater than t_o, between which we assume the image-space location of the platform as the last known location in the image, overriding the requirement that $\sigma_V > \sigma_o$.

3.3 Moving Platform Tracking

A simple PID-based controller is applied to achieve the robust flight and control activities of the drone. Both velocity, as well as position error, is resolved while descending towards the moving vehicle. The parameters of the controller are tuned accordingly. For the smooth and perfect landing of the drone on a moving vehicle all the above steps are needed and these are stitched together into a robust algorithm.

Algorithm 1 depicts the whole process starting from detection of the moving vehicle applying the optical flow and contour on image feed from rgb camera. Further, the centroid of the vehicle is calculated using the corners from the obtained vehicle contour. Using the visual servo control, the velocity component v_{xy} and v_z of moving vehicle relative to UAV is calculated. After this, a basic PID controller is applied to resolve the positional error as well. When UAV reaches a certain threshold the straight vertical landing sequence automatically starts and disarms the UAV on the landing platform.

Algorithm 1 Landing of UAV on moving platform

STEP-1: Moving Vehicle detection:
Using Gunnar-Farneback optical flow algorithm
Optical Flow Equation : $\frac{dI}{dx}u + \frac{dI}{dy}v + \frac{dI}{dt} = 0$
Assume,constant pixel intensities between the two frames
STEP-2: Segmentation of moving vehicle:
OpenCV BGR to HSV
STEP-3: Find landing point
find contour of the vehicle and mark the centroid
then transform from camera coordinate to world coordinate
i.e. Ground Vehicle pose (X_v, Y_v, Z_v)
STEP-4: Visual Servoing:
If $\sigma_V < \sigma_o$, the UAV slowly moves upwards
else $\sigma_V > \sigma_o$, landing site came into view
then Get V_{xy} and V_z
STEP-5:PID Controller:
Drone pose $(X_d = 0, Y_d = 0, Z_d)$
Calculate the difference in all axis$X_{dif}, Y_{dif}, Z_{dif}$
1. Drone will follow the ground Vehicle based on PID on (X_{dif}, Y_{dif})
2. Landing control $\propto 1/(X_{dif}, Y_{dif})$
then start descending towards ground Vehicle
If $X_{dif} < 0.1, Y_{dif} < 0.1, Z_{dif} < 0.5$
then Land on the platform
Else Repeat STEP-2

4 Simulation

A gazebo-based 3D simulation environment is used to evaluate and validate of the proposed method as shown in Fig. 1. The complete setup is built on Ubuntu 18.04 LTS with ROS (Robot Operating System) as the main framework. PX4-based flight control software is used to establish SITL simulation, which uses MAVLink protocol for parallel communication. For the Arial vehicle, an Iris drone with a downward facing RGB camera publishing images at 20 Hz with a resolution of 640×480 placed at the bottom, and a husky with an enlarged top plate for a large landing surface is used. Husky is used because it is rugged and an outdoor-ready unmanned ground vehicle (UGV) which is suitable for research prototyping applications.

Fig. 1 Optical flow-based moving vehicle detection on Gazebo simulator

The navigation of Husky is controlled by using the Husky navigation stack with a simple PID controller. The whole method is tested by moving the Husky in different directions and at different speeds. The autonomous landing on the center of a landing platform of the moving vehicle travels at a speed of 10 m/s with an error of around 20 cm. One of the test results of trajectories of the drone as well as husky is shown in Fig. 2 In order to perform the task, the different segments mentioned above need to perform seamlessly without any lag. Here, ROS plays an important role to manage all the nodes and enables smooth communication between them. The communication between nodes and topics is represented in Fig. 3.

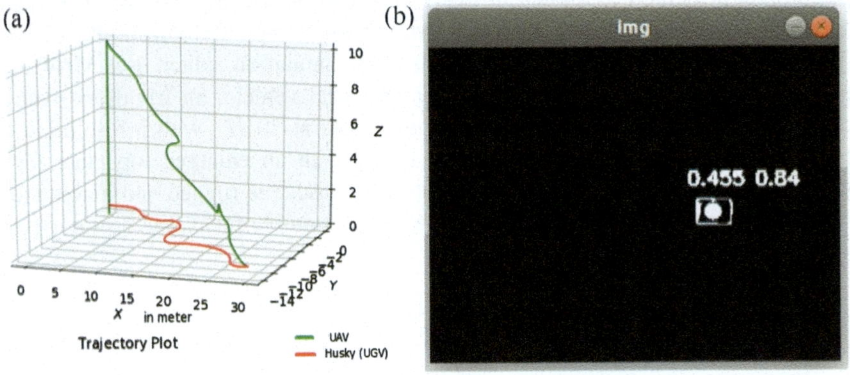

Fig. 2 **a** The trajectory of UAV and UGV. **b** Pose detection based on the contour of the vehicle

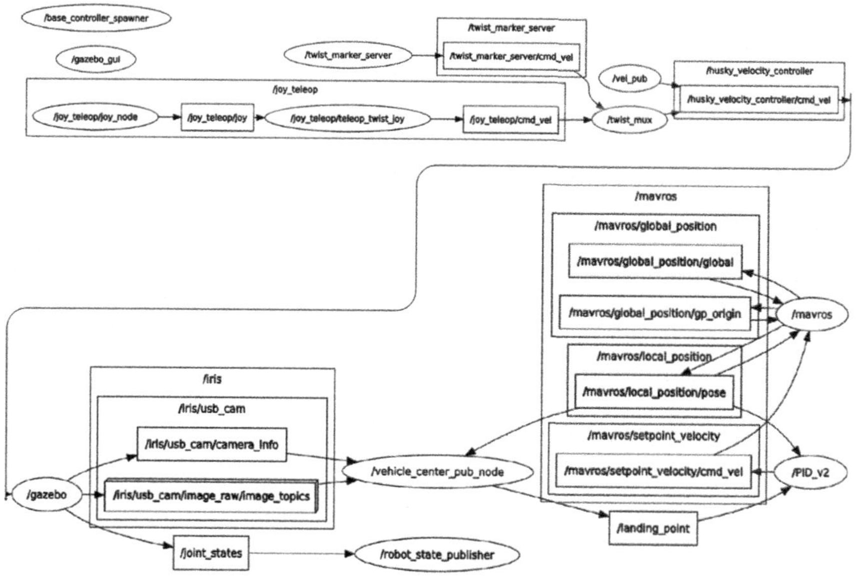

Fig. 3 RQT graph depicting the ROS nodes and data flow within the implemented code

5 Conclusion

In this work, we have introduced a different approach to landing a drone autonomously on a general moving platform. The method includes a detection stage where a moving vehicle is identified using optical flow and the contour of the vehicle is drawn to get the center of the landing platform. In the next stage, visual servo control is used to match the velocity of both the UAV and UGV which is Iris drone and husky here respectively. A PID controller is used finally to handle the position and velocity errors in order to achieve an unruffled landing maneuver. For safety and validation, the method is tested in Gazebo virtual world. The approach is computationally inexpensive and more generalized than the available marker-based methods. The presented work achieved a satisfactory precision level in the simulation environment, which can be further tested in a real-time environment.

Acknowledgements The work has been financially supported by i-Hub Innovation Foundation (IHFC), New Delhi. The authors are grateful for the support and facility provided by CSIR-CEERI, Pilani.

References

1. Polvara R, Sharma S, Wan J, Manning A, Sutton R (2018) Vision-based autonomous landing of a quadrotor on the perturbed deck of an unmanned surface vehicle. Drones. 2(2):15
2. Daud SM, Yusof MY, Heo CC, Khoo LS, Singh MK, Mahmood MS, Nawawi H (2022) Applications of drone in disaster management: a scoping review. Sci Justice 62(1)
3. Zhang Y, Yu Y, Jia S, Wang X (2017) Autonomous landing on ground target of UAV by using image-based visual servo control. In: 2017 36th Chinese control conference (CCC), 2017, pp 11204–11209
4. Kishore K, Dalai S, Jangir Y, Singh S, Rohan M, Shashank D, Katta SS, Saha SK (2022) 3D pure pursuit guidance of drones for autonomous precision landing. In: 2022 13th Asian control conference (ASCC), pp 2218–2222
5. Chen S, Zhao X, Sun Z, Xiang F, Zhixin S (2019) Shape recognition with recurrent neural network. In: International conference on artificial intelligence and security. ICAIS 2019, pp 341–350
6. Gautam A, Singh M, Sujit PB, Saripalli S Autonomous quadcopter landing on a moving target. Sensors 22(3):1116
7. Baca T, Stepan P, Saska M (2017) Autonomous landing on a moving car with unmanned aerial vehicle. In: 2017 European conference on mobile robots (ECMR)
8. Lebedev I, Erashov A, Shabanova A (2020) Accurate autonomous UAV landing using vision-based detection of ArUco-marker ICR 2020. In: Lecture notes in computer science, vol 12336. Springer
9. Huang X, Xu Q, Wang J Vision-based autonomous landing of UAV on moving platform using a new marker conference series: materials science and engineering
10. Rodriguez-Ramos A, Sampedro C, Bavle H et al (2019) A deep reinforcement learning strategy for UAV autonomous landing on a moving platform. J Intell Robot Syst 93:351–366
11. Xie J, Peng X, Wang H, Niu W, Zheng X (2020) UAV autonomous tracking and landing based on deep reinforcement learning strategy. Sensors (Basel) 20(19):5630 Oct 1
12. Feng Y, Zhang C, Baek S, Rawashdeh S, Mohammadi A (2018) Autonomous landing of a UAV on a moving platform using model predictive control. Drones 2(4):34
13. Borowczyk A, Nguyen DT, Nguyen AP, Nguyen DQ, Saussié D, LeNy J (2017) Autonomous landing of a multirotor micro air vehicle on a high velocity ground vehicle. In: 2017, 20th IFAC world congress, pp 10488–10494
14. Kumar G, Pandya H, Gaud A, Krishna KM (2017) Pose induction for visual servoing to a novel object instance. In: IEEE/RSJ international conference on intelligent robots and systems (IROS) 2017:2953–2959
15. Bradski G The OpenCV library. Dr Dobb's J Softw Tools
16. Harris CR, Millman KJ, van der Walt SJ, Gommers R, Virtanen P, Cournapeau D (2020) Array programming with NumPy. Nature 585:357–62

Innovative Two-Axle Vehicle with Improved Ride Comfort via Blended Active Vibration Control

Rocco Libero Giossi, Rickard Persson, and Sebastian Stichel

Abstract A mechatronic two-axle rail vehicle with only one suspension step is introduced in the Shift2Rail project Pivot2. This vehicle design reduces the vehicle weight in comparison to standard bogie vehicles. However, having only one suspension step drastically decreases passenger comfort. Thus, hydraulic actuators are introduced instead of passive dampers and active modal sky-hook control is applied. Due to the strong interaction between the running gear frame and the carbody, a blended modal solution is applied where a percentage of the acceleration of the frame is used in the feedback loop in addition to the acceleration of the carbody. To assess the performance of the controllers, simulations are carried out with the vehicle running at constant speeds from 10 km/h to 120 km/h on tangent track with high level of track irregularities. First, multiplicative dimensional reduction method (M-DRM) sensitivity analysis is applied to determine the importance of the control variables and subsequently a genetic algorithm (GA) optimization is performed to identify the control gains for each speed. The blended control proposed here can improve passenger comfort with respect to a standard modal control while maintaining similar energy and force usage.

Keywords Two-axle vehicle · Modal control · Blended control · Comfort evaluation

1 Introduction

In the Shift2Rail project Pivot2, an innovative passenger vehicle with single axle running gear and one suspension step is proposed, drastically reducing the vehicle weight. The concept also features active suspension to overcome the deficiency in vibration attenuation the omitted suspension step brings. With the proposed design, both investment and maintenance costs are reduced, which provides a significantly

R. L. Giossi (✉) · R. Persson · S. Stichel
KTH Royal Institute of Technology, Stockholm 10044, Sweden
e-mail: roccolg@kth.se

S. K. Saha and K. Rama Krishna (eds.), *Recent Developments in Multibody Dynamics*, Lecture Notes in Mechanical Engineering,
https://doi.org/10.1007/978-981-97-7525-5_9

improved business case for the introduction of active suspension in rail vehicles. A composite material frame will be used both as structural and suspension element.

Different approaches are possible to actively improve passenger comfort [5]. Among those, a major distinction between semi-active and active suspension may be done. Semi-active suspensions show significative improvement when the initial comfort of the vehicle is sufficiently close to an acceptable limit. Improvement of the comfort using magnetic rheologic dampers was shown in field tests by Kim et al. in [9]. However, when the improvement required by the active system is massive, fully active systems must be applied. In these cases, the actuators generally replace standard dampers (Pacchioni et al. [11] and Qazizadeh et al. [12]). For the innovative two-axle vehicle, the latter is to be considered. In fact, for the vehicle equipped with standard dampers, the weighted acceleration in vertical direction is $0.42\,\mathrm{m/s^2}$ at 100 km/h where the limit for a comfortable ride according to the EN12299 standard [4] is $0.25\,\mathrm{m/s^2}$.

In terms of control strategies, modal sky-hook control, being among one of the most simple and effective strategies, finds application in field tests (Qazizadeh et al.[12] and Sugahara et al.[13]). Modal control is also considered here as a starting point for the development of the controller for the proposed vehicle. Preliminary simulation results on tangent track with irregularities show that the modal sky-hook approach is sufficient to reach an admissible level of passenger comfort. Nevertheless, due to the strong interaction between the running gear frame and the carbody, the performance can be further improved by taking a percentage of the frame acceleration as control variable. A similar concept was introduced by Alujević et al. in [1, 2] for a 3 d.o.f. system, showing improvements with respect to a standard sky-hook control. The concept proposed by Alujević is modified to fit the purpose of the modal controller selected for our case. As for the 3 d.o.f. system, the blended control performs better than the standard modal control.

In the following, the innovative vehicle proposed in the Pivot2 project is presented together with the actuation system used. Subsequently, the methodology for control design and the control system will be described. The simulation results are discussed from comfort, force and energy perspectives and are summarized in conclusions.

2 Vehicle and Actuation Modelling

The innovative solution proposed here is composed of two main components. The first one being the innovative vehicle itself and the second one is the actuators used to control the vibration in the carbody. As mentioned above, the single suspension step makes it necessary to have an active control to improve the passenger ride comfort.

The vehicle is meant to be an alternative to the existing Metro Madrid class 8000 vehicles with the objective of significantly reducing the tare weight per meter. The expected savings are about 400 kg/m [6, 8].

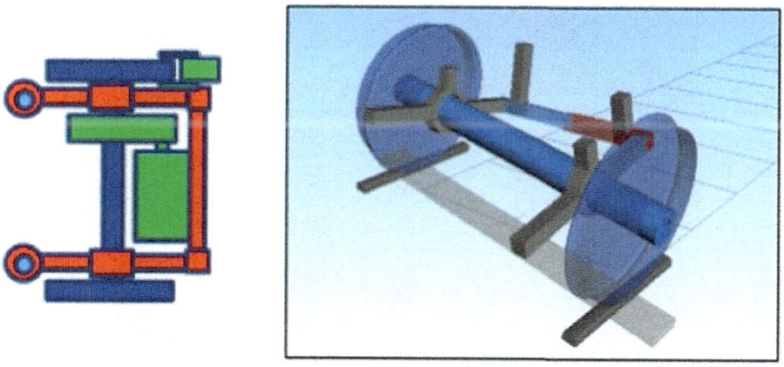

Fig. 1 Running gear (left) and SIMPACK implementation (right) [7]

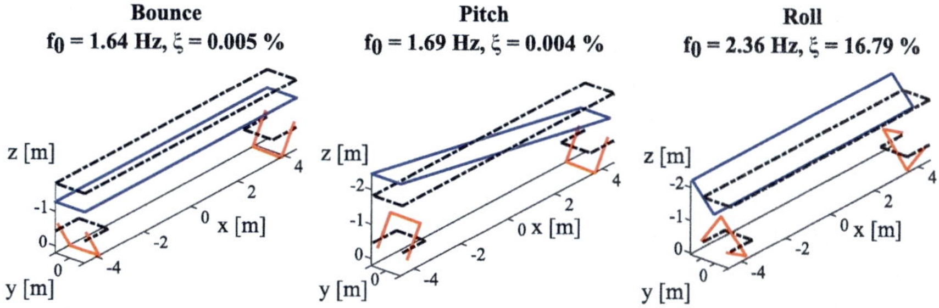

Fig. 2 Innovative vehicle first eigenmodes: Bounce (left), Pitch (centre) and Roll (right)

The innovative vehicle is modelled in SIMPACK where special attention is given to the connection element between the wheelsets and the carbody. A schematic representation of the U-shaped connection frame is shown in Fig. 1 (left). The connection frame will be manufactured in composite material giving the possibility to integrate the anti-roll bar within itself. The composite material frame is simplified with a simple but effective representation. To simulate the anti-roll bar, the required torsional stiffness (250 kNm/rad) connects two rigid halves of the frame. In this way, the movement shown in Fig. 1 (right) is achieved.

Despite the importance of flexible carbody modes in the evaluation of passenger comfort, in this work, due to the absence of a physical model of the carbody, a rigid body with a mass of 13 tons corresponding to the tare weight is considered. Once the model of the vehicle has been set in SIMPACK, it is possible to calculate the eigenmodes of the system. In Fig. 2, the most important eigenmodes (bounce, pitch, and roll) for this study are represented. The eigenfrequencies and the non-dimensional damping ratios for each mode are given.

Close attention must be paid to the movement of the end extremities of the U-shaped frame. In fact, in all the shown mode shapes, the end extremities move in counterphase with respect to the carbody respective points. Due to the position of the actuators (Fig. 3 (right)), these motions can create limitations in the control action and thus decrease the algorithms effectiveness in reducing carbody vibrations.

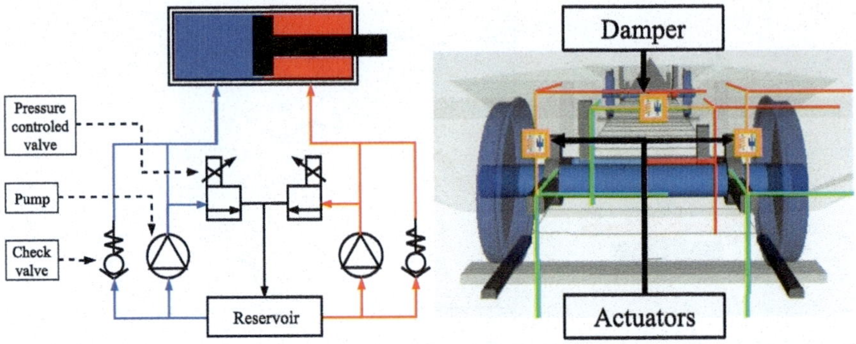

Fig. 3 Actuator schematic (left) and actuator position in SIMPACK model (right)

The actuator is modelled in MATLAB/Simulink to increase the trustworthiness of the results achieved with the study. The actuator scheme (Fig. 3 (left)) shows the two chambers of the double acting hydraulic cylinder where each chamber is controlled by a separate circuit powered by a fixed displacement pump. In each circuit, the pressure is controlled through a pressure-controlled valve that can control the pressure proportionally to an input current. A complete description of the modelled actuator can be found in [8].

3 Methodology to Design the Control Algorithm

Due to the aforementioned problem of counterphase movement of the actuator attachment points, two control algorithms are compared. The first one is the most standard modal control algorithm where the carbody acceleration signals (co-located with the actuatorâŁ™s actuation points) are used as feedback signals for a modal sky-hook controller. The second one, called blended controller, takes advantage of the additional measurement of acceleration from the end extremities of the U-shaped frame to reflect a more complete behaviour of the vehicle before feeding it back to the modal sky-hook controller. A sensitivity analysis is first done to ensure that the control parameters used are effective. Subsequently, the two control algorithms are optimized with the same optimization procedure. In this way, a fair comparison between the effectiveness of each approach can be performed. Here, a genetic algorithm (GA) is used since a clear relationship between control gains and comfort is not known.

The comfort is evaluated in vertical direction with the EN12299 standard [4] with the vehicles running at constant speed on a 1000 m tangent track with ERRI_High track irregularities [3]. The vehicle speeds are chosen to be from 10 km/h to the allowable speed of 120 km/h with an interval of 10 km/h. Both the sensitivity analysis and the optimization procedure are applied to each simulation scenario.

4 Control System

As mentioned above, different control approaches can be applied to increase the performance in terms of passenger comfort. First, the control algorithms are described and subsequently the sensitivity analysis and its results are presented.

4.1 Control Algorithm

At first, a standard modal control is implemented. This control requires the knowledge of the mode shapes of the system. The accelerations measured in the carbody are combined through the mode shape matrix $[\phi]$ and are subsequently integrated by a bandpass filter to obtain the modal velocity feedback signals to be used as control variables for the sky-hook controller. Only the first three rigid body modes, which are shown in Fig. 2, are considered, i.e., bounce, pitch, and roll. Three control gains, k_1, k_2, and k_3 are used. The signals are then recombined to generate the input currents to be provided to the pressure-controlled valves. The input currents are thus defined as

$$\underline{i}_{valve} = -[\phi][K][\phi]^T \int \ddot{\underline{y}} dt, \tag{1}$$

where, $[\phi]$ is the mode shape matrix, $[K]$ is the diagonal gain matrix (k_1, k_2, and k_3), and $\ddot{\underline{y}}$ is the carbody accelerations vector. Moreover, $\int \cdot$ represents the combination of the integration function (1/s) and a high-pass filter with cut-off frequency of 1 rad/s to disregard quasi-static behaviour.

As mentioned before, the control described above does not consider the influence of the support at which the actuators are placed, i.e., the U-shaped frame. The concept studied by Alujević et al. in [1, 2] for a 3 d.o.f. systems that showed performance improvements compared to a standard sky-hook approach, have here been adopted to the modal sky-hook control. Here, a percentage of the modal contribution of the frame accelerations is added to the input for the modal sky-hook controller. The actuators are acting in opposite direction to the measured frame accelerations. Thus, the frame modal contributions must be subtracted from the carbody ones to avoid positive feedback in the frame actuation points. In this blended case the input currents are defined,

$$\underline{i}_{valve} = -[\phi][K]\left([\phi]^T \int \ddot{\underline{y}} dt - [\alpha][\phi^f]^T \int \ddot{\underline{y}}^f dt\right), \tag{2}$$

where, the superscript f stands for the frame contribution and the diagonal matrix $[\alpha]$ is the blended control matrix. The modal blended gains (α_1, α_2, and α_3) can vary between 0 and 1, representing the percentage of the frame modal contribution considered in the control procedure.

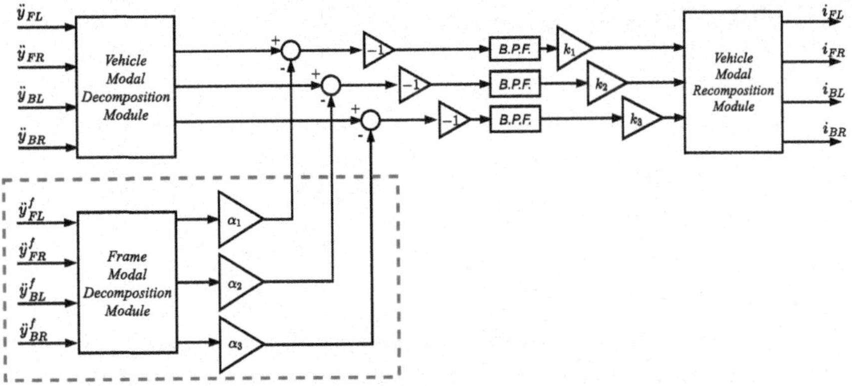

Fig. 4 Control scheme (where subscript F stands for front, B for back, L left and R right)

In Fig. 4, the two control algorithms are shown. The addition or removal of the red dashed block lets the controller change from standard modal control to blended modal control.

4.2 Sensitivity Analysis

Before the optimization is performed, both control schemes undergo a sensitivity analysis to sort the importance of control parameters. The multiplicative dimensional reduction method (M-DRM) [14] is applied to perform the sensitivity analysis. The method was applied successfully in railway applications in [10]. The method is extremely useful when high number of parameters is involved. In fact, to investigate N parameter each one with M possible values, the number of simulations required is only N*M. Other methods may require more simulations up to a maximum value of N^M for an exhaustive study. The method can suffer of inaccuracy if the number of values M for each parameter is small. Nevertheless, as shown in [14], a general trend can be identified even with a small number M.

The M-DRM has been applied to both controllers for each considered speed separately with 15 values per considered control parameter. The objective function of the sensitivity analysis is the vertical comfort evaluated with the EN12299 standard. In Fig. 5, the results obtained from the sensitivity analysis are summarized for both modal and blended control. Each column represents the cumulative sensitivity in percentage per vehicle speed. The bars inside each column show the shared importance of each parameter at that speed. Thus, taking as example the modal control sensitivity for the front part of the vehicle (Fig. 5 (top-left)) at 100 km/h, the bounce control parameter k_1 has about 25% influence over the comfort while the pitch parameter k_2 has 75% and the roll parameter k_3 0%.

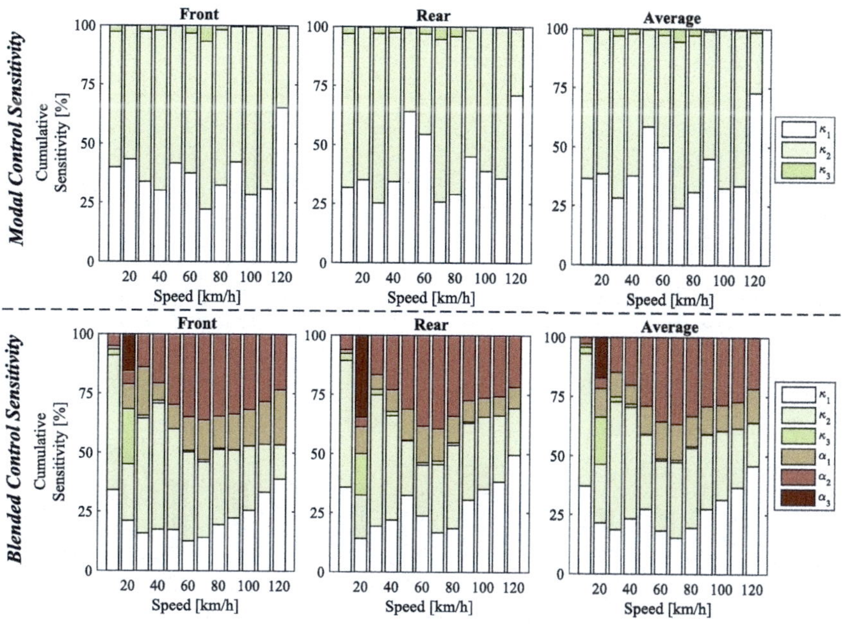

Fig. 5 Sensitivity analysis results: modal control results (top) and blended control results (bottom)

The major outcome of the sensitivity analysis is the absence of the roll control (both modal gain k_3 and blended gain α_3) from the significant control parameters. This can be explained by the presence of a passive damper in lateral direction (Fig. 3 (Right)) that provides a non-dimensional damping ratio of 17% for the roll motion.

Thus, following the results of the sensitivity analysis, the GA optimization is applied to the bounce and pitch control parameters only. The fit function is, in this case, the mean value between the comfort values between the front and rear part of the vehicle measured at the floor level according to the EN12299 standard.

5 Results

Since the control optimization is applied to the vehicle running at constant speed, specific control gains are found for each case. The obtained results in terms of control gains are shown in Fig. 6. In a real case application, a look-up table and interpolation may be implemented with the optimal control gains for any speed. It is possible to see that the control gains increase when the vehicle speed decreases, especially for the modal control. The trend can be explained by carbody vibration amplitudes which increase with speed. Thus, the control gains should be low at high speed and high at low speed if comparable forces would be produced. In contrast, the blended

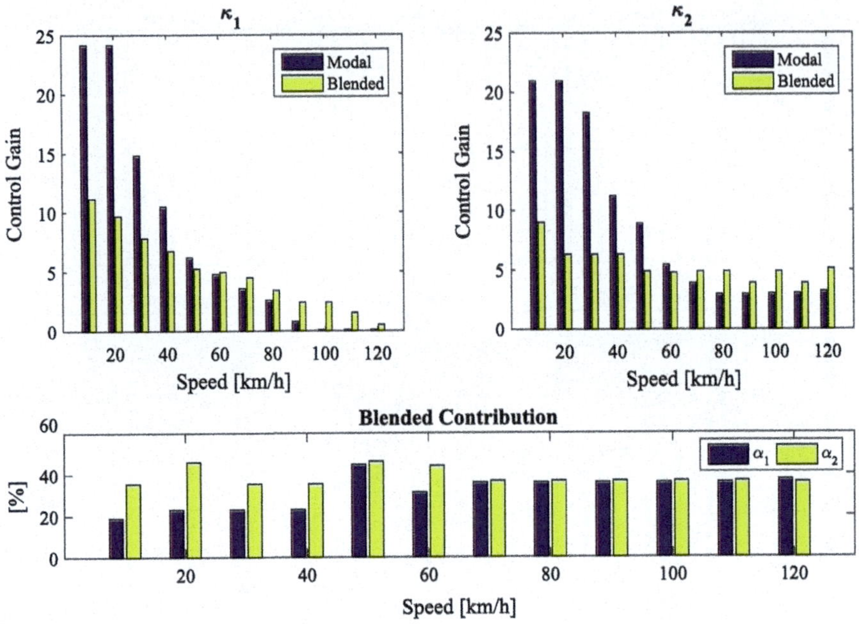

Fig. 6 Control gains versus vehicle speed: k_1 (top-left), k_2 (top-right) and blended contribution (bottom)

contributions α_1 and α_2 are less dependent on vehicle speed reaching a stable point around 35% above 70 km/h.

The comfort evaluation resulting from the optimization procedure for both controllers is given in Fig. 7. It is possible to notice that the comfort with the blended controller never exceeds the comfortable limit given in the EN12299 standard, while the modal controller reaches this limit for the vehicle top speed. Nevertheless, both algorithms can maintain the ride comfort in the comfortable zone throughout the vehicle operating speed range. A more detailed comparison between the two controllers can be done by defining an improvement factor as,

$$\Delta(a) = \frac{a_{modal} - a_{blended}}{a_{modal}} \cdot 100, \tag{3}$$

being a_{modal} the weighted acceleration for the modal control respectively $a_{blended}$ the one for the blended control. The improvement defined by Eq. (3) shows that the blended controller provides a general improvement to the system behaviour (Fig. 7 (bottom)).

A worsening in comparison to the modal performance is seen for very low speeds (10 and 20 km/h) where the absolute weighted accelerations anyway are extremely low and well below the acceptable comfort limit. In an average sense, the blended

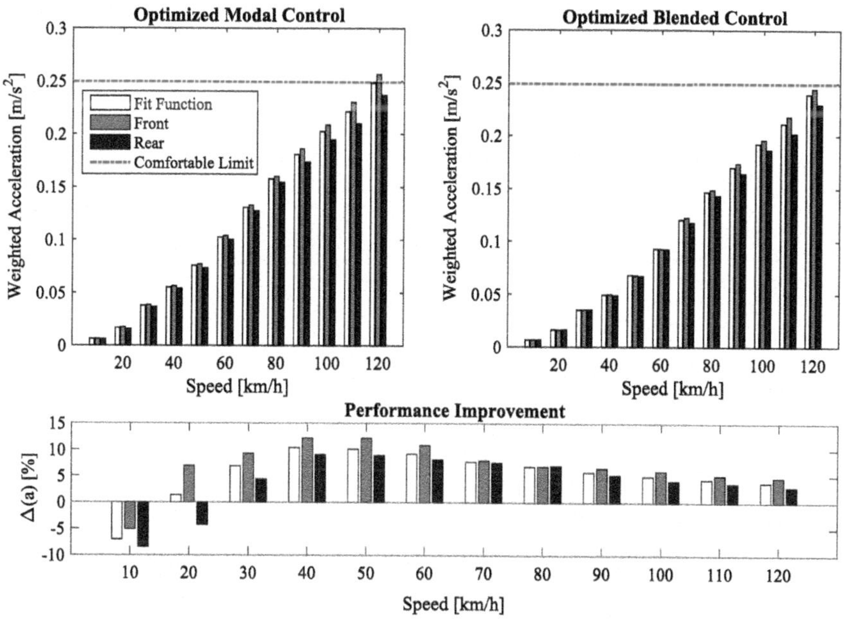

Fig. 7 Comfort results: modal control (top-left), blended control (top-right) and their comparison (bottom)

controller provides an improvement of 5.4%, with 6.9% in the front part of the vehicle and 3.9% in the rear one.

Despite the effectiveness of the proposed blended control with regard to ride comfort improvement, other aspects must be taken into consideration. These are the maximum used force, the actuator displacement and the RMS power. In Fig. 8, these quantities are shown comparing the modal and with the blended control as function of vehicle speed. Here, the maximum force per each speed among the four actuators is considered. The same is done for the actuator displacement, while the sum of the four actuators for the RMS power is used. For a better comparison of the results, Eq. (3) is used once again to evaluate the improvement of the blended control in comparison to the modal one. The blended control generally performs better for vehicle speeds lower than 60 km/h. The benefit of blended control is less clear above 60 km/h with cases with higher force, larger actuator displacement and more power used.

A more in-depth analysis can be done to compare the performances of the two controllers. Despite the alternate performance in terms of relative force, displacement, and rms power in dependence of the vehicle speed, a relevant measure is to compare the maximum of these parameters independently from the vehicle speed at which they occur. Additionally, the energy consumed by the actuators throughout a trip on Metro Madrid line 10 can be determined. In fact, thanks to the known velocity distribution in line 10 (Fig. 8 (bottom-right)), it is possible to weight the energy

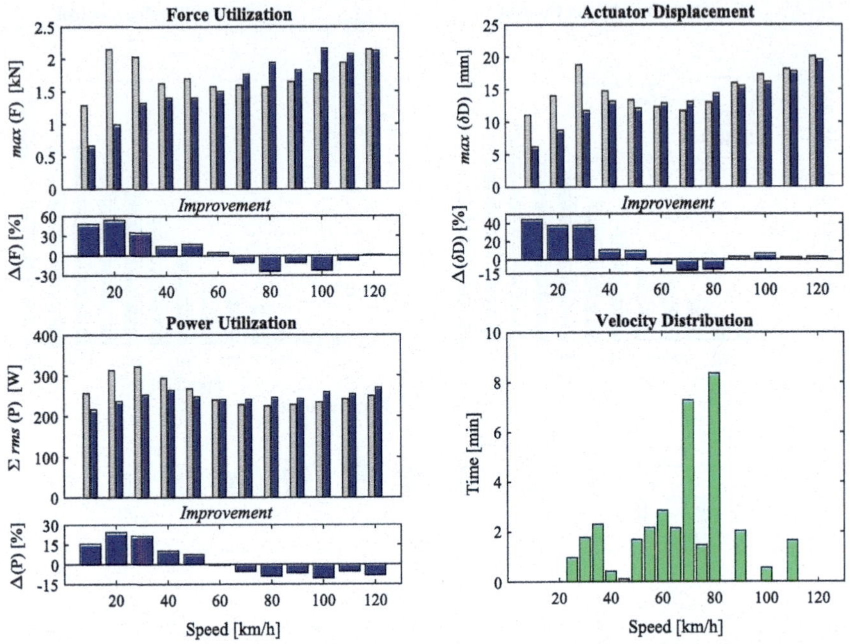

Fig. 8 Force utilization (top-left), actuator displacement (top-right), power utilization (bottom-left) and Metro Madrid line 10 velocity distribution (bottom-right). Modal control in grey and blended control in cyan

Table 1 Comparison between force, displacement, RMS power and energy consumption between modal and blended control

Control type	Max force [kN]	Max disp. [mm]	Max power [W]	Energy cons. [kJ]
Modal	2.15 (@120 km/h)	20.1 (@120 km/h)	83.9 (@30 km/h)	495.8
Blended	2.16 (@100 km/h)	19.5 (@120 km/h)	68.1 (@120 km/h)	493.5

obtained during the simulations to match the ones expected in the real case scenario per each vehicle speed,

$$\hat{E}^v = \sum_{j=1}^{4} \int_0^{t_{sim}} P_j^v dt \cdot \frac{t_{track}}{t_{sim}}. \tag{4}$$

Here, the power used during the simulations is integrated over the simulations time and subsequently weighted by the ratio between the time spent on that speed during one trip on line 10 (t_{track}) and the simulation time (t_{sim}). Once summed for every vehicle speed, the estimation of the energy consumption for each controller over line 10 can be achieved. The results given in Table 1 show that the overall performance of the two controllers is similar, and it is not sufficient to judge the benefit of one controller with respect to the other.

6 Discussion

The comfort evaluation shows that active vertical suspension significantly improves ride comfort making it possible to introduce a vehicle with single axle running gears and only one suspension step instead of conventional bogies, promising substantial weight, and cost savings.

Blended control improves the passenger comfort further compared to the standard modal approach at the expense of a more complex system. This involves more measurement sensors and more knowledge of the vehicle behaviour, i.e., the frame mode shapes. Therefore, the ride comfort benefit should be weighed against the extra initial and maintenance cost of the blended control when selecting the type of control.

7 Conclusion

An innovative two-axle vehicle with only one suspension step has been presented. Due to the poor ride comfort because of the omitted second suspension, step active control in vertical direction has been introduced to improve passenger comfort together with an actuation system for this purpose.

Two controllers are compared, a standard modal controller and the blended controller introduced here. To verify the effectiveness of the two controllers, both undergo the same procedure. First, a M-DRM sensitivity analysis is done, showing that only bounce and pitch motion influence the control performance. Secondly, a GA optimization is used to find the control gains. The two steps are repeated for each considered vehicle speed with the vehicle running at constant speed from 10 km/h to 120 km/h with a velocity span of 10 km/h on 1000 m tangent track with ERRI_High track irregularities.

Results show that the blended control improves the passenger comfort by around 5.4% compared to the modal controller, while it maintains comparable actuator force, displacement, RMS power and energy consumption.

Acknowledgements This project has received funding from the Shift2Rail Joint Undertaking (JU) under grant agreement 881807. JU receives support from the European Union's Horizon 2020 research and innovation programme and the Shift2Rail JU members other than the Union.

References

1. Alujević N, Wolf H, Gardonio P, Tomac I (2011) Stability and performance limits for active vibration isolation using blended velocity feedback. J Sound Vib 330(21):4981–4997. https://doi.org/10.1016/j.jsv.2011.05.020
2. Alujević N, Wolf H, Depraetere B, Zhao G, Domazet Z, Pluymers B, Desmet W (2015) Self-tuneable velocity feedback for active isolation of random vibrations in subcritical two degree of freedom systems. Acta Acust United Acust 101(5):950–963. https://doi.org/10.3813/AAA.918890
3. Bergander B, Kunnes W (1993) ERRI B176/DT 290: B176/3 benchmark problem. Results and assessment. Technical report, European Rail Research Institute
4. CEN; EN 12299: railway applications—ride comfort for passengers—measurement and evaluation (2009)
5. Fu B, Giossi RL, Persson R, Stichel S, Bruni S, Goodall R (2020) Active suspension in railway vehicles: a literature survey. Railw Eng Sci. https://doi.org/10.1007/s40534-020-00207-w,http://link.springer.com/10.1007/s40534-020-00207-w
6. Giossi RL, Persson R, Stichel S (2020) Improved curving performance of an innovative two-axle vehicle: a reasonable feedforward active steering approach. Veh Syst Dyn 1–24. https://doi.org/10.1080/00423114.2020.1823005, https://doi.org/00423114.2020.1823005
7. Giossi RL, Persson R, Stichel S (2020) Gain scaling for active wheelset steering on innovative two-axle vehicle. In: Klomp M, Bruzelius F, Nielsen J, Hillemyr A (eds) Advances in dynamics of vehicles on roads and tracks. Springer International Publishing, Cham, pp 57–66
8. Innovative running gear solutions for new dependable, sustainable, intelligent and comfortable rail vehicles, deliverable 3.2—new actuation systems for conventional vehicles and an innovative concept for a two-axle vehicle. Technical report (2019)
9. Kim HC, Shin YJ, You W, Jung KC, Oh JS, Choi SB (2017) A ride quality evaluation of a semi-active railway vehicle suspension system with MR damper: railway field tests. Proc Inst Mech Eng, Part F: J Rail Rapid Transit 231(3):306–316. https://doi.org/10.1177/0954409716629706
10. Mousavi-Bideleh SM, Berbyuk V (2016) Multiobjective optimisation of bogie suspension to boost speed on curves. Veh Syst Dyn 54(1):69–96. https://doi.org/10.1080/00423114.2015.1114655
11. Pacchioni A, Goodall RM, Bruni S (2010) Active suspension for a two-axle railway vehicle. Veh Syst Dyn 48(sup1):105–120. https://doi.org/10.1080/00423111003668252, http://www.tandfonline.com/doi/abs/10.1080/00423111003668252
12. Qazizadeh A, Persson R, Stichel S (2015) Preparation and execution of on-track tests with active vertical secondary suspension. Int J Railw Technol 4(1):29–46. https://doi.org/10.4203/ijrt.4.1.2, http://www.ctresources.info/ijrt/paper.html?id=77
13. Sugahara Y, Kazato A, Koganei R, Sampei M, Nakaura S (2009) Suppression of vertical bending and rigid-body-mode vibration in railway vehicle car body by primary and secondary suspension control: Results of simulations and running tests using Shinkansen vehicle. Proc Inst Mech Eng Part F: J Rail Rapid Transit 223(6):517–531. https://doi.org/10.1243/09544097JRRT265
14. Zhang X, Pandey MD (2014) An effective approximation for variance-based global sensitivity analysis. Reliab Eng Syst Saf 121:164–174. https://doi.org/10.1016/j.ress.2013.07.010, http://dx.doi.org/10.1016/j.ress.2013.07.010

A Linear Frequency Domain Solver Workflow for Fast Simulation of Transmission Systems

K. D. Bauer, J. Haslinger, and G. Offner

Abstract The contribution presents a complete frequency solver workflow applied to automotive transmissions modeled as multi-body systems consisting of mechanical components like rotating rigid and elastic shafts interconnected by gear contacts and supported by bearing joints. Solving the equations of motion in frequency domain based on a linearized model yields the periodic steady-state results very fast compared to transient time integration methods, where fast oscillating components may decay slowly. The frequency domain solver workflow is described in detail from getting a loaded state of the model, which is used for linearization, up to solving linear equation systems for each non-negligible frequency load component. The presented solver workflow is applied to a simple gearbox model. Resulting vibrations from the linear frequency domain solution are compared against results of a transient time domain solution, where the frequency domain solution matches well the time domain results but is obtained within a fraction of CPU time.

Keywords Frequency domain · Transmission systems · Linearized equation of motion

K. D. Bauer · J. Haslinger (✉)
Radon Institute for Computational and Applied Mathematics, Austrian Academy of Sciences and MathConsult GmbH, Altenberger Str. 69, 4040 Linz, Austria
e-mail: josef.haslinger@mathconsult.co.at

K. D. Bauer
e-mail: kdbauer@mathconsult.co.at

G. Offner
Advanced Simulation Technologies, AVL List GmbH, Hans-List-Platz 1, 8020 Graz, Austria
e-mail: guenter.offner@avl.com

S. K. Saha and K. Rama Krishna (eds.), *Recent Developments in Multibody Dynamics*, Lecture Notes in Mechanical Engineering,
https://doi.org/10.1007/978-981-97-7525-5_10

1 Introduction

When key design parameters are varied in the early design phase of transmission systems, a fast computation of the dynamic properties is essential, e.g. [14, 15]. This is especially true for automotive transmissions modeled as multi-body systems consisting of mechanical components like rotating rigid and elastic shafts interconnected by joints like gear contacts and supported by bearing joints.

A standard transient time integration for such systems based on the BDF integration method for a floating frame of reference approach is described in [3]. In case of simulating typical application cases like periodic steady-state motion, the computation may take minutes up to several hours depending on the model complexity, as transient oscillations from fast moving components may decay slowly. Solving the equations of motion in frequency domain based on a linearized model yields the steady-state results within seconds.

The main motivation for a frequency domain solver (FDS) is to get a "fast" solution compared to the standard time domain solver (TDS). An FDS operates on the same model depth as is used for the regular transient calculation in time domain. As the FDS is operating on a linearized version of the model, this method is best suited for models, where nonlinear effects do not play a dominant role. Otherwise, the FDS may not yield meaningful results.

In general, the linearization of the nonlinear equations of motion of a multi-body model leads to a second-order ODE with time-varying coefficient matrices [6, 12]. Here, the focus is on the description of the workflow how to obtain a fast solution: it is assumed that the first-order Taylor approximation leads to a second-order ODE with constant, time-independent coefficient matrices [4, 5]. For instance in transmission applications, time-dependent variable meshing effects in gear meshes are not represented by the linear FDS as presented here. They have to be taken into account as external loads by, e.g. an additional "tooth contact analysis" (TCA).

This paper is organized as follows: in Sect. 2, the basic equations of motion of multi-body systems, their linearization, and their solution are summarized from the theoretical point of view. Section 3 presents the complete frequency solver workflow: starting from the setup of the multi-body model, which is analogous to the one in a TDS, calculation of initial velocities, kinetostatics, linearization of the model around the kinetostatics configuration to the solution of the linear FDS equations for determining frequency result components. The presented solver workflow is applied to a simple gearbox model in Sect. 4. As the assumption on the coefficient matrices is rather strictly ones, an outlook how to extend the FDS workflow to a more general class of models is given in Sect. 5.

2 Basic Equations of Motion, Their Linearization, and Their Solution

2.1 *Nonlinear Equations of Motion in Floating Frame of Reference*

The time-dependent description of the motion of the multi-body system is based on the floating frame of reference approach. This approach leads to the standard system of nonlinear second-order differential algebraic equations as used in the multi-body simulation software AVL EXCITE^{TM} [3]. The differential equation of this system for body i is given as:

$$\boldsymbol{M}_i^{FEM}\boldsymbol{q}_i''+\boldsymbol{D}_i^{FEM}\boldsymbol{q}_i' + \boldsymbol{K}_i^{FEM}\boldsymbol{q}_i + \\ + \ \boldsymbol{f}_i^{inertia}(\boldsymbol{y}_i, \boldsymbol{y}_i', \boldsymbol{y}_i'') + \sum_m \boldsymbol{f}_i^{joint_m}(\tilde{\boldsymbol{y}}, \tilde{\boldsymbol{y}}') = \boldsymbol{f}_i^{load}(\boldsymbol{y}_i, \boldsymbol{y}_i') \quad (1)$$

with state variables $\boldsymbol{y}_i$, including global variables and the local, elastic variables $\boldsymbol{q}_i$. $\boldsymbol{M}_i^{FEM}$, $\boldsymbol{D}_i^{FEM}$ and $\boldsymbol{K}_i^{FEM}$ denote the mass, damping and stiffness matrix, resulting from spatial discretization. The spatial finite element discretization can be obtained by beam mass models or by optionally condensed 3D-volume models, see [13]. Nonlinear inertia forces—like Coriolis and gyroscopic forces and torques—are collected in $\boldsymbol{f}_i^{inertia}$. External forces and moments applied at a single body are covered by $\boldsymbol{f}_i^{load}$. Forces and moments resulting from coupling between the bodies are contained in $\boldsymbol{f}_i^{joint}$. The interaction between different bodies—including gear contact—is modeled by force elements (joints) and not by kinematic constraints. The joint forces/moments do not only depend on coordinates of the body itself but also on possibly all states of other bodies, which are summarized in the vector $\tilde{\boldsymbol{y}}$. The equations of motion are completed by algebraic constraints such as reference conditions for uniquely splitting global and local motion parts for an elastic body.

The equations of motion form an implicit second-order differential algebraic equations system (DAE) with differentiation index two [2]. For the computation of the dynamic behavior of the multi-body system, the DAE system (1) has to be solved for each body for both global and local motion quantities. For the time integration of the DAE system, a Backward Differentiation Formula (BDF) scheme is used. As the bodies are connected by joint force laws, iterative decoupling of the full multi-body system into a sequence of smaller units on body level is applied for efficiency reasons. In each iteration step, the joint forces are kept fixed. After the body iteration step is finished, the joint forces are updated and predicted for the next iteration step of the bodies. For further details, see [3].

2.2 Linearization of Equations of Motion

In their most general form, the equations of motion of the overall multi-body system with n_b bodies can be summarized as:

$$\boldsymbol{f} = \left(\boldsymbol{f}_1^T, \boldsymbol{f}_2^T, \cdots, \boldsymbol{f}_{n_b}^T\right)^T = \boldsymbol{f}^{load} \tag{2}$$

where $\boldsymbol{f}$ contains all single body terms $\boldsymbol{f}_i$ on the left hand side of (1). Similarly $\boldsymbol{f}^{load}$ contains all the load components of all single bodies.

A standard linearization procedure (first-order Taylor approximation) is used to linearize the equations of motion about a reference trajectory $\boldsymbol{y}_0(t)$ of the multi-body system such that the solution $\boldsymbol{y}(t) = \boldsymbol{y}_0(t) + \boldsymbol{\Delta y}$ is represented as small perturbations about the reference solution. The following equations are obtained:

$$\begin{aligned} \boldsymbol{f}(\boldsymbol{y}_0''+\boldsymbol{\Delta y}'', \boldsymbol{y}_0' + \boldsymbol{\Delta y}', \boldsymbol{y}_0 + \boldsymbol{\Delta y}) &\approx \boldsymbol{f}(\boldsymbol{y}_0'', \boldsymbol{y}_0', \boldsymbol{y}_0)+ \\ &+\frac{\partial \boldsymbol{f}}{\partial \boldsymbol{y}''}|_{(\boldsymbol{y}_0'', \boldsymbol{y}_0', \boldsymbol{y}_0)} \boldsymbol{\Delta y}'' + \frac{\partial \boldsymbol{f}}{\partial \boldsymbol{y}'}|_{(\boldsymbol{y}_0'', \boldsymbol{y}_0', \boldsymbol{y}_0)} \boldsymbol{\Delta y}' + \frac{\partial \boldsymbol{f}}{\partial \boldsymbol{y}}|_{(\boldsymbol{y}_0'', \boldsymbol{y}_0', \boldsymbol{y}_0)} \boldsymbol{\Delta y} \end{aligned} \tag{3}$$

These equations are rewritten in the usual matrix form of the linearized equations of motion:

$$\boldsymbol{M}(t)\boldsymbol{\Delta y}'' + \boldsymbol{D}(t)\boldsymbol{\Delta y}' + \boldsymbol{K}(t)\boldsymbol{\Delta y} = \boldsymbol{f}^{load}(\boldsymbol{y}_0) - \boldsymbol{f}(\boldsymbol{y}_0'', \boldsymbol{y}_0', \boldsymbol{y}_0) \tag{4}$$

where $\boldsymbol{M}$, $\boldsymbol{D}$ and $\boldsymbol{K}$ denote the—in general time-dependent—mass, damping and stiffness matrix of the linearized multi-body system with:

$$\boldsymbol{M} = \left.\frac{\partial \boldsymbol{f}}{\partial \boldsymbol{y}''}\right|_{(\boldsymbol{y}_0'', \boldsymbol{y}_0', \boldsymbol{y}_0)} \qquad \boldsymbol{D} = \left.\frac{\partial \boldsymbol{f}}{\partial \boldsymbol{y}'}\right|_{(\boldsymbol{y}_0'', \boldsymbol{y}_0', \boldsymbol{y}_0)} \qquad \boldsymbol{K} = \left.\frac{\partial \boldsymbol{f}}{\partial \boldsymbol{y}}\right|_{(\boldsymbol{y}_0'', \boldsymbol{y}_0', \boldsymbol{y}_0)} \tag{5}$$

Most parts of the linearization of the equations of motion (1) are computable analytically. While **M** describes bodies, **D** and **K** describe both bodies and joints. Linearizations include the FEM-based body mass, damping and stiffness matrices in case of an elastic body. Additional gyroscopic terms may result from linearization of the inertia forces $\boldsymbol{f}^{inertia}$ such that

$$\boldsymbol{D} = \boldsymbol{D}_i^{FEM} + \boldsymbol{G} \tag{6}$$

with $\boldsymbol{G}$ representing the gyroscopic matrix depending on the angular velocities of the bodies. Analogously the linearized stiffness matrix $\boldsymbol{K}$ may contain beneath the elastic stiffness matrix $\boldsymbol{K}_i^{FEM}$ components that depend on angular velocities of bodies.

The linearization of the joint forces/moments $\boldsymbol{f}^{joint}$ yields contributions to the linearized stiffness and damping matrices like in the force equilibrium approach in [8]. For a joint connecting one node on body i and one node on body j, the 12×12 joint stiffness matrix $\boldsymbol{K}^{joint}$ is assembled from four 6×6 blocks as [7]:

$$\boldsymbol{K}^{joint} = \begin{pmatrix} \boldsymbol{K}_{ii}^{joint} & \boldsymbol{K}_{ij}^{joint} \\ \boldsymbol{K}_{ji}^{joint} & \boldsymbol{K}_{jj}^{joint} \end{pmatrix}. \tag{7}$$

If the joint connects a flexible body with another flexible body, the partial derivatives are directly added to the body stiffness matrix at the positions of the connected nodes. The 6×6 blocks on the diagonal are added to the overall stiffness matrix at the diagonal positions corresponding to body i, node k and body j, node l. The off-diagonal 6×6 blocks are added at block positions (body i, node k / body j, node l) and vice versa to the overall stiffness matrix.

2.3 Solution of Linearized Equations of Motion in Frequency Domain

The linearized equations (4) may be solved efficiently. In general, the matrices in (4) do depend on time; they form a linear time-variant ordinary differential equation system. Most often the matrices are time periodic. If additionally, the external loads are periodic, solution methods are proposed in [6, 12, 16] for modal analysis. An iterative algorithm, especially for powertrain simulation is proposed in [1].

In the following, we restrict ourselves to models where the linearization of the equations of motion do lead to a time-invariant differential equation, i.e. the matrices $\boldsymbol{M}$, $\boldsymbol{D}$ and $\boldsymbol{K}$ do not depend on time:

$$\boldsymbol{M}\boldsymbol{\Delta y}'' + \boldsymbol{D}\boldsymbol{\Delta y}' + \boldsymbol{K}\boldsymbol{\Delta y} = \boldsymbol{f}^{load}(\boldsymbol{y}_0) - \boldsymbol{f}(\boldsymbol{y}_0'', \boldsymbol{y}_0', \boldsymbol{y}_0) \tag{8}$$

Furthermore, the external load along the reference trajectory $\boldsymbol{f}^{load}(\boldsymbol{y}_0(t))$ shall be periodic with primitive period T.

Any periodic function $f(t)$ can be represented as a complex-valued Fourier series

$$f(t) = \sum_{n=-\infty}^{\infty} f_n e^{i\omega_n t} \tag{9}$$

on a frequency grid $\omega_n = n\Delta\omega = n\frac{2\pi}{T}$ where T is the (primitive) period of the function. In practice, the amplitudes $f_n \in \mathbb{C}$ are obtained by some form of discrete Fourier transform (DFT), such as FFT.

All the described matrices as well as the vectors $\mathbf{q}$ (displacement vector, or vector of generalized coordinates) and $\mathbf{f}$ (active load vector, or harmonic excitation vector) contain real-valued elements. The constraint, that all functions are expected to be real-valued, corresponds to a symmetry

$$f_n = f_{-n}^* \tag{10}$$

between the positive- and negative-frequency components, which allows to skip the calculation of negative-frequency Fourier coefficients. The time-dependent function is then reconstructed as

$$f(t) = f_0 + \sum_{n=1}^{\infty} (f_n e^{i\omega_n t} + f_n^* e^{i\omega_{-n} t}) = f_0 + \sum_{n=1}^{\infty} 2Re(f_n e^{i\omega_n t}) \tag{11}$$

We are interested in steady-state solutions $\boldsymbol{\Delta y}(t)$ under application of a periodic load $\mathbf{f}(t)$. Substituting the ansatz

$$\boldsymbol{\Delta y}(t) = \sum_n \boldsymbol{\Delta y}_n e^{i\omega_n t} \tag{12}$$

in the equation of motion yields

$$\sum_n \left[-\omega_n^2 \mathbf{M} + i\omega_n \mathbf{D} + \mathbf{K}\right] \boldsymbol{\Delta y}_n e^{i\omega_n t} = \sum_n \mathbf{f}_n e^{i\omega_n t} \tag{13}$$

Since the functions $e^{i\omega_n t}$ are linearly independent, and since we do not need to calculate negative-frequency coefficients separately, we thus obtain the equations for $n \geq 0$:

$$\left[-\omega_n^2 \mathbf{M} + i\omega_n \mathbf{D} + \mathbf{K}\right] \boldsymbol{\Delta y}_n = \mathbf{f}_n \tag{14}$$

The linear complex-valued equation system (14) with the system matrix

$$-\omega_n^2 \mathbf{M} + i\omega_n \mathbf{D} + \mathbf{K} \tag{15}$$

often called *dynamic stiffness matrix* is solved then separately for each of the excitation frequencies. The components of the excitation $\mathbf{f}_n$ are determined using the harmonic analysis: discrete Fourier transform (DFT)—see, for example [15], or fast Fourier transform (FFT).

3 Workflow for a Linear Frequency Domain Solution

3.1 General FDS Workflow

The multi-body model of the transmission system is set up in the same way as in a standard transient time integration. All rigid and flexible body types may be used. The class of joint types is restricted to the ones that may be linearized analytically or by using finite differences. As in standard time calculation, initial velocities of the bodies are calculated or corrected. Especially angular velocities of bodies are

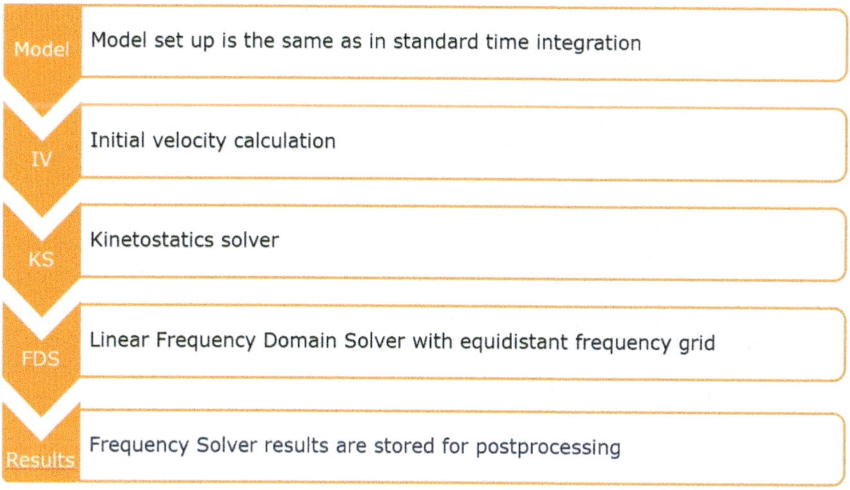

Fig. 1 Schematic representation of the workflow

computed as the rotation of the bodies around a fixed axis constitutes the reference motion.

Then, using the kinetostatics solver [10], a loaded configuration of the model is sought for, i.e., external loads are applied and a static equilibrium position is computed resulting in pre-loaded bodies and joints, yielding in particular gear contacts being closed.

This equilibrium position is the starting point for the linearization of the model, where effects from the rotation are included. The joint types are linearized by the method of finite differences or—in cases of relatively simple force laws—analytically [7]. The usual differential equation (8) of second order with constant coefficient matrices resulting from the linearization describes the small motion of the model around the linearization point, where constant rotation of the individual bodies around a fixed axis is assumed. This differential equation is then solved in frequency domain. On equidistant frequency grids, the external loads are first transformed to frequency domain by FFT. The dynamic stiffness matrix is generated and the corresponding linear equation system is solved for each non-negligible frequency load component. The results can be represented in the frequency domain as well as recomputed in the time domain.

The workflow is schematically represented in Fig. 1.

3.2 *Detailed Description of FDS Workflow*

After the transmission model is created, the following steps are carried out consecutively in a linear FDS-based solution procedure:

1. Calculation of initial velocities Initial global velocities of the bodies of the multibody system are calculated (or corrected) in the same way as in standard transient integration [3]. The knowledge of consistent initial velocities is essential as the (uniform) rotational motion of the bodies with the calculated angular velocities is used as reference motion. In transmission systems, usually the angular velocity of one or some shafts are prescribed, but not all global initial velocities are known. Therefore, it is necessary to compute those missing taking into account constraints from joint interactions. Hence, the joint force laws are temporarily replaced by corresponding algebraic constraints and the velocities are computed by taking into account the Jacobian of the constraints.
2. Kinetostatics for a loaded configuration External loads are applied and a static equilibrium position is computed. In particular, an appropriate contact is established, for example, for gear joints.

For a transient time integration, the kinetostatic state represents the pre-loaded conditions at the starting time point. The reduction of initial oscillations due to inaccurate initial conditions but also the determination of the correct start configuration, which avoids nonphysical penetrations of structural components, are the main issues in time domain.

In FDS, the kinetostatic solution is used as the state where the model is linearized. Together with a possibly (uniform) rotation from the initial value calculation, it defines the reference trajectory of a body.

For the kinetostatic state, a solution $\boldsymbol{y}_{ks} = \boldsymbol{y}_0(0)$ of the right-hand side of (8) is sought for [10] such that

$$\boldsymbol{f}^{load}(\boldsymbol{y}_{ks}) - \boldsymbol{f}(0, 0, \boldsymbol{y}_{ks}) = \mathbf{0}. \tag{16}$$

Inertia effects are neglected as the initial accelerations and velocities—especially of the flexible components of a body—are not known and are assumed to be zero. In principle, stiffness and joint forces/moments are brought into equilibrium with the applied external loads.

The numerical solution of (16) is based on a modified Quasi-Newton method. As the Jacobian of (16) is not regular, in general, a regularization term is added based on the mass/inertia properties of the bodies.
3. Linearization At the kinetostatics equilibrium configuration $\boldsymbol{y}_{ks}$, the linearization of the model is carried out with the computation of the linearized matrices, i.e. mass-, damping-, stiffness matrices including gyroscopic parts from rotation—if requested. For details, see Sect. 2.2.
4. Computation of frequency components of external loads External loads $\boldsymbol{f}^{load}$ are specified in time domain, whether directly by the user, or precomputed by other components of the simulation suite. They are decomposed into their Fourier components by FFT, using separate frequency grids with the frequency resolution derived from the primitive period of each load item, and the upper frequency limit specified by the user. The results are then transferred to a global frequency grid with a frequency resolution specified by the user. In order to avoid introducing a beat in the signal, the

frequency components are binned into the closest frequency of the global frequency grid, rather than distributed to the two closest frequencies.

This *multi-grid* sampling approach has arisen from the need to handle artifacts of the FFT. Originally, the user specified a single equidistant frequency grid, defined by an upper frequency limit and a frequency resolution. Frequency components that deviated from this grid by as little as 10^{-13} parts of the frequency resolution would produce windowing artifacts, that cause a full occupation of the grid, severely reducing the performance of the subsequent solution step, and introducing high-frequency error terms into calculated velocities. Avoiding this would require to find the smallest common period of all force terms, and using its inverse as the frequency resolution, which is unrealistically difficult for real-world models. The high sensitivity to small rounding errors proved to make internal adjustment of the frequency resolution unreliable.

Additionally, this would still leave the issue of aliasing artifacts, which caused occupation of unoccupied frequencies. While more benign than broadening of the spectrum from windowing artifacts, aliasing artifacts would still produce high-frequency error terms, that were enhanced in derivatives. The multi-grid load sampling approach removes windowing artifacts completely and ensures that aliasing artifacts occur only as small contributions to already-occupied frequency components, rather than introducing *new* frequency components.

5. Linear frequency domain solution The linearized equations of motion are solved. The solution represents small amplitude periodic steady-state motions around the reference motion including constant rotation of individual bodies around a fixed axis (in absolute coordinate frame).

The dynamic stiffness matrix (15) for each non-zero external load frequency component is set up and the corresponding complex-valued linear system (14) is solved. The result is the complex amplitude of motion at this frequency of each DOF of the system.

The solution of the linear systems (14) is carried out in $O(N_\omega N_{DOF}^2)$ operations, where N_{DOF} represents the number of DOFs in the model and N_ω the number of occupied frequencies in the global frequency grid. The solution is the cheaper the more load frequency components are zero on the global frequency grid.

6. Storage of results and postprocessing Native FDS results are stored for postprocessing. Additionally, the re-computation (synthesis) of FDS results in time domain may be requested.

4 Application to a Gearbox Model

The presented solver workflow is applied to a simple gearbox model as depicted in Fig. 2. In the application example, three elastic gear shafts (the green input shaft, the blue layshaft, and the yellow output shaft) supported by in total five deep groove ball bearing joints in a rigid housing are connected by two cylindrical gear joints. Additionally, the input shaft supports the output shaft through a radial bearing joint.

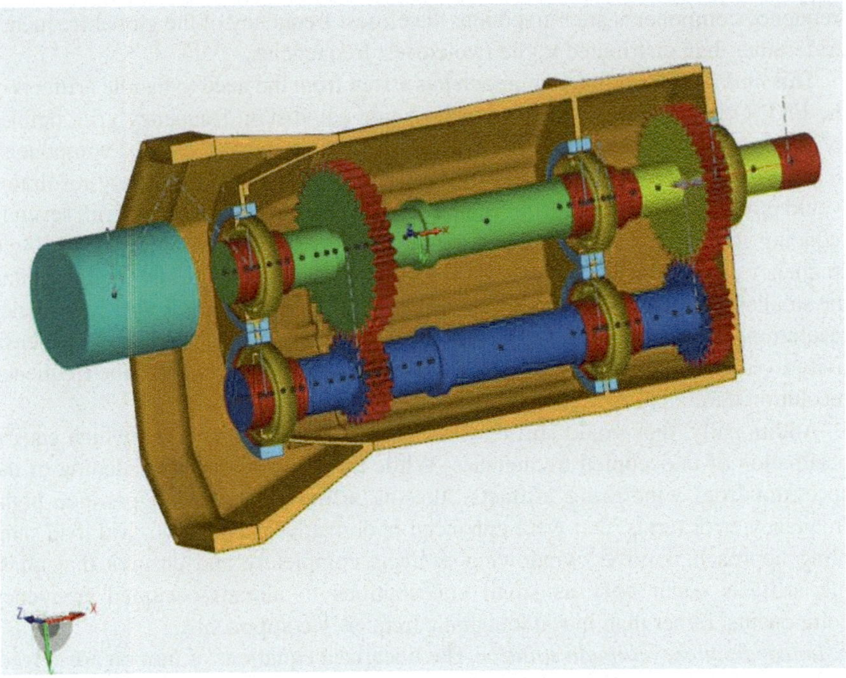

Fig. 2 The model of a simple gearbox with three elastic gear shafts: input shaft (green), layshaft (blue), output shaft (yellow) supported by ball bearings in a rigid housing and connected by cylindrical gears

In total, the model consists of 426 elastic DOFs (input shaft: 168 DOFs, layshaft: 204 DOFs, output shaft: 54 DOFs).

In general, the principal tasks that need to be processed by a gear joint evaluation are: contact detection, constitution of the deformation field in the gear mesh, force distribution in the mesh and—if requested—calculation of friction and damping forces. The modeling of the gear meshing process within a multi-body system reveals a wide variety of possible approaches [11]. In this model, the two cylindrical gear joints are parameterized in such a way that they are linear: both mesh stiffness and damping values of the cylindrical gear joints assumed to be constant, and tooth separation is not considered. The ratio of teeth of the gear joint connecting input shaft and layshaft is 38:19 and that of the gear joint connecting layshaft and output shaft is 27:39.

Deep groove ball bearings are typically applied for the bearings of gearshafts. They show non-linear behavior, as the bearings are configured such that periodic variations of the resultant bearing stiffness are taken into account. The radial bearing joint connecting input and output shaft is a linear one.

Angular motion of the input shaft is realized through predefined motion of the cylinder (turquoise) on the left side connected by a linear rotational coupling joint. An oscillating moment is applied to the output shaft where the period of the oscillation coincides with one rotation of the output shaft.

The FDS workflow is applied to the model with a predefined motion of the cylinder of 3000 rotations per minute (rpm):

1. The initial velocity calculation gives rotational speeds of 3000 rpm for the input shaft, 6000 rpm for the layshaft (in negative direction of rotation of the input shaft), and about 4153.846 rpm (= 69.231 rps) for the output shaft. Clearly, the rotation ratio of the shafts is determined by the properties of the two gear joints.
2. Next step is the kinetostatic solution which converges within 75 iterations. The angular moment applied at the output shaft leads to slightly distorted shafts. Additionally, the nodes of the shafts are translationally displaced according to force equilibrium conditions which are mainly determined by the two gear joints.
3. The resulting kinetostatic state is used as configuration for the linearization of the model. All components of the model are linearized analytically except for the five ball bearings and the two gear joints. The partial derivatives of the joint forces with regard to node positions and velocities are approximated by the method of finite differences. As the gear joints are linear, the results of numerical linearization may be checked against analytical derivation in literature, e.g. [9].
4. Next step is the computation of frequency components of the applied external moment. FFT is carried out on a load-specific frequency grid. The only significant frequency component of the external moment is calculated to be 69.231 Hz (equal to rotation velocity of output shaft). Very small numerical artifacts appear at 28 additional multiples of the base frequency of 69.231 Hz. A non-zero frequency component is computed for the constant part of the moment at 0 Hz.
5. The global equidistant frequency grid for the frequency domain solver itself is defined with lower frequency limit 0 Hz, upper limit 5000 Hz and frequency resolution of 0.833333 Hz. The main external load component of the applied moment is binned to the closest frequency, which is 69.167 Hz on the global grid. Similarly, the components resulting from numerical FFT artifacts are binned to their closest frequencies. Out of a number of 6001 possible frequency grid points, the linear equation system (14) for determining solution components is solved for 30 of them.

The resulting vibrations from a linear frequency domain solution are compared against results of a complete time domain solution up to 2000 deg reference angle, see Fig. 3 and for a detailed view Fig. 4. As the gear joints couple both transverse and torsional motion of the three shafts, the position of one transverse component of the gear joint connection node at the left side of the layshaft is compared as an example. The additional oscillations in the TDS solution in Fig. 4 are due to nonlinear behavior of deep groove ball bearings, but the basic behavior of the solution is well reproduced.

The time domain solution is accepted to be reliable, but it takes much more simulation time before a stationary state is reached: the time domain solution takes

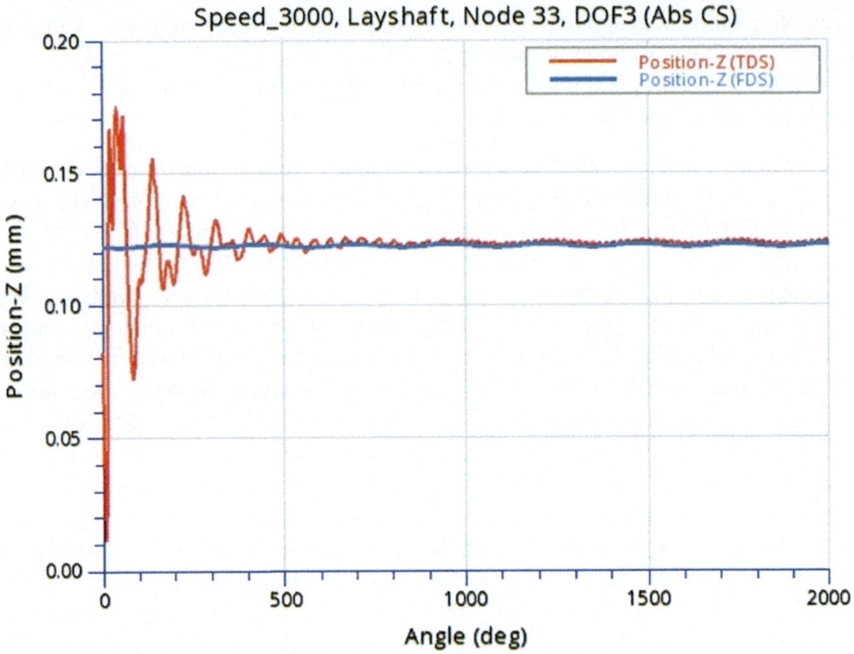

Fig. 3 Comparison of motion results obtained from a time domain (TDS, red) and frequency domain (FDS, blue) solver. The steady-state motion is calculated immediately by FDS

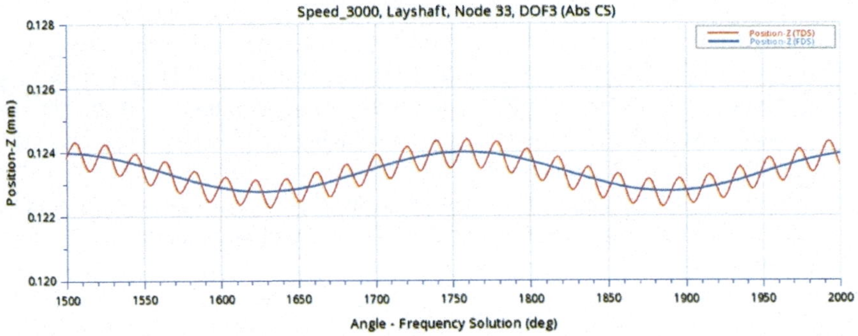

Fig. 4 Details of Fig. 3 from 1500 to 2000 deg rotation angle. The additional oscillations in the TDS solution are due to nonlinear behavior of the ball bearings, but the basic behavior of the solution is well reproduced

115.4 s CPU time, compared to only 7.9 s (1.5 s for native FDS results and 6.4 s for synthesis in time domain) in the presented frequency domain solver workflow.

5 Conclusion and Outlook

A linear frequency domain solver workflow has been presented and applied to a simple transmission system. The restriction to linear ordinary differential equations of second order with time-independent coefficient matrices limits the approach to systems with nearly negligible nonlinear effects. However, the method is very efficient to get a fast overview of the dynamic properties of simple systems.

In a next step, the method will be extended to include, e.g. gear meshing effects. In a preprocessing step - the so-called "tooth contact analysis" (TCA) - the contact forces in the mesh are calculated. Then, these forces are applied as external loads in the linear frequency domain solver workflow.

Alternatively, gear meshing effects may be taken into account as stiffness fluctuations contributing to a time-varying stiffness matrix. An iterative frequency domain approach with alternating computations in frequency and time domain is in development [1].

Acknowledgements Part of the work of Klaus-Dieter Bauer and Josef Haslinger was supported by the Government of Upper Austria within the "FIT Struktur" framework and by the FFG within the project "FFG DigiLab".

References

1. Bauer KD, Haslinger J, Offner G, Parikyan T (2022) Approaches for going beyond linear frequency domain powertrain simulation. In: Proceedings of ECMI 2021
2. Drab CB (2007) Dynamic simulation of power train related systems. Ph.D. thesis, Johannes Kepler Universität, Linz
3. Drab CB, Engl HW, Haslinger JR, Offner G, Pfau RU, Zulehner W (2009) Dynamic simulation of crankshaft multibody systems. Multibody Syst Dyn 22:133–144
4. Friswell M, Penny J, Garvey S, Lees A (2010) Dynamics of rotating machines. Cambridge aerospace series. Cambridge University Press
5. Genta G (2007) Dynamics of rotating systems. Mechanical engineering series. Springer, New York
6. Han S, Bauchau O (2020) Simulation and stability analysis of periodic flexible multibody systems. Multibody Syst Dyn 50(4):381–413
7. Haslinger J, Offner G, Sopouch M, Zinkiewicz B (2021) Modal analysis of planetary gearsets. In: Proceedings of the NAFEMS world congress: Salzburg, Austria. NAFEMS
8. Kang JS, Bae S, Lee JM, Tak TO (2003) Force equilibrium approach for linearization of constrained mechanical system dynamics. J Mech Des 125:143–149
9. Kaplan JA, Dousti S, Allaire PE, Nichols BR, Dimond TW, Untaroiu A (2013) Rotor dynamic modeling of gears and geared systems. In: Volume 7A: structures and dynamics. American Society of Mechanical Engineers
10. Lorenz N, Offner G, Colla A (2013) Preloaded initial state evaluation of flexible multibody systems. In: ECCOMAS thematic conference on multibody dynamics 2013. Zagreb, pp 463–472
11. Neusser Z, Sopouch M, Schaffner T, Priebsch HH (2010) Multi-body dynamics based gear mesh model for prediction of gear dynamics and transmission error. In: Transmission and driveline 2010, SAE world congress & exhibition

12. Nguyen VK, Nguyen PD, Hoang MC (2009) Linearization and parametric vibration analysis of some applied problems in multibody systems. Multibody Syst Dyn 22(2):163–180
13. Offner G (2011) Modelling of condensed flexible bodies considering non-linear inertia effects resulting from gross motions. Proc IMechE Part K: J Multi-body Dyn 225:204–219
14. Parikyan T, Bukovnik S (2019) Turbocharger dynamic analysis: concept-phase simulation in frequency domain. In: SIRM 2019, Copenhagen, pp 893–902
15. Parikyan T, Resch T (2012) Statically indeterminate main bearing load calculation in frequency domain for usage in early concept phase. In: ASME 2012 internal combustion engine division fall technical conference
16. Pedersen R, Santos IF, Hede IA (2010) Advantages and drawbacks of applying periodic time-variant modal analysis to spur gear dynamics. Mech Syst Signal Process 24(5):1495–1508

Multibody Dynamic Study of Subassembly Transfer Flask Under Seismic Excitation

Sasidhar Inakollu, S. D. Sajish, and Jose Varghese

Abstract The proposed fuel handling scheme of the future Indian fast reactors (FBR 1 and 2) envisages the use of a dedicated flask called subassembly transfer flask (STF) for handling of subassemblies out of the reactor vessel. The STF transfers the fuel subassembly (SA) between fuel transfer port (FTP-I) on roof slab and fuel transfer position 2 (FTP-2) on RCB floor and vice versa. This machine consists of a flask mounted over a self-propelled carriage, which moves between the fuel transfer port of reactor vessel at one end and fuel transfer port of inert gas filled SA transfer carriage cell (STC cell) at another end. STF transfer/receives transfer pot holding irradiated/FSA (Fresh Subassembly) to/from the transfer carriage through the transfer ports. STF traverse between the two ports on rails and raises/lowers the transfer pot by wire ropes, Multibody dynamics approach is used to study the dynamic behavior of the STF under an earthquake loading. The present work is a detailed MBD investigation of the performance of STF under seismic events to evaluate the forces experienced by the different components of the machine and to study the motion of SA inside STF during a seismic excitation. Rigid body dynamic behaviour is considered for the docked STF at FTP loaded with SA, but the locking pins are modelled as flexible elements. Nonlinear contact behaviour between STF and rails during seismic excitation is observed and the functionality effectiveness of locking mechanism provided for the machine is verified through MBD investigation using RecurDyn solver for analysis.

S. Inakollu (✉) · S. D. Sajish · J. Varghese
Reactor Design and Technology Group, Indira Gandhi Centre for Atomic Research, Kalpakkam, Tamilnadu, India
e-mail: sasidhar99@igcar.gov.in

S. D. Sajish
e-mail: sajish@igcar.gov.in

J. Varghese
e-mail: varghese@igcar.gov.in

S. K. Saha and K. Rama Krishna (eds.), *Recent Developments in Multibody Dynamics*, Lecture Notes in Mechanical Engineering,
https://doi.org/10.1007/978-981-97-7525-5_11

1 Introduction

During fuel handling, STF lifts spent SA in sodium filled transfer pot (TP) from the reactor vessel using Offset Handling Machine (OHM) mounted over the roof slab through FTP-I, moves to the FTP-2 and deposits the TP on a trolley working inside an inert cell and vice versa, for fresh SA handling. For raising and lowering the TP inside STF, a dedicated hoist called gripper hoist is provided, which raises and lowers the STF Gripper, which in turn holds TP, at FTP-1/FTP-2. The dropping of core component during handling is a possible scenario, hence single failure proof design is considered for gripper hoist. The hoist consists of two wire ropes connected to the top of the gripper assembly on one end, and to a rope drum through a set of pulleys on the other end. The rope drum is supported over an axle driven by a geared motor on one end. A brake is provided on the other end of the rope drum shaft through a gearbox. The geared motor is also having an inbuilt brake. A tension sensing mechanism is provided between the main pulley and rope drum to make sure the loading/unloading of subassembly is done before undocking. The hoisting mechanism is supported by support structure with main support columns laterally stiffened by structural steel angles, forming a rectangular cage structure as shown in Fig. 1 which supports upper flask body and transfer the loads to carriage assembly. The central canal of flask body that houses the fuel subassembly and SA Gripper is appropriately shielded with Grey Cast Iron blocks to reduce the nuclear radiation emitted out by the spent subassembly.

Fig. 1 Subassembly transfer flask

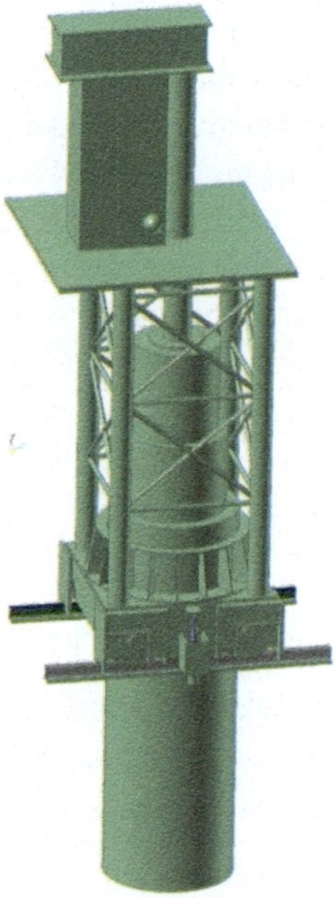

STF traverses between FTP-1 and FTP-2 on rails. It has a coupling mechanism to form a leak-tight lock with the Floor Valve above FTP for safe SA handling due to sodium present on the outer surface of TP. Buffers are provided at the top of FTP locations to bring the STF to rest completely, and then the locking pin is latched to the locking nut to provide complete positional stability during fuel handling operation. The shielding block is supported on the carriage by a skirt support. The carriage assembly consists of wheels and geared motors by which the STF moves on the rails, four seismic arrestors are attached to the carriage assembly to avoid derailing and four anti-toppling guide bars are provided to prevent the machine from toppling in case of a seismic event.

In this study, the response spectrum available at the support elevation [1] is used to generate the time history data to simulate seismic excitation. The displacement time history data generated are given as the input excitation to the rails on which the STF is resting. All the components of STF are considered rigid except the locking pin, which must deform to create reaction forces from the carriage structure during excitation. Contacts are given from the rails to four wheels and seismic anchors, locking pin to the hole and carriage structure. The contact force time histories between the wheels and rail are generated to estimate the loss of contact due to vertical component of the seismic excitation. The STF is allowed to rest completely on the rails in the initial condition of study and after the reaction forces are stable. The seismic motion for 20 s is applied to study the dynamic behavior of the machine under seismic excitation (Fig. 2).

Fig. 2 Internals of STF

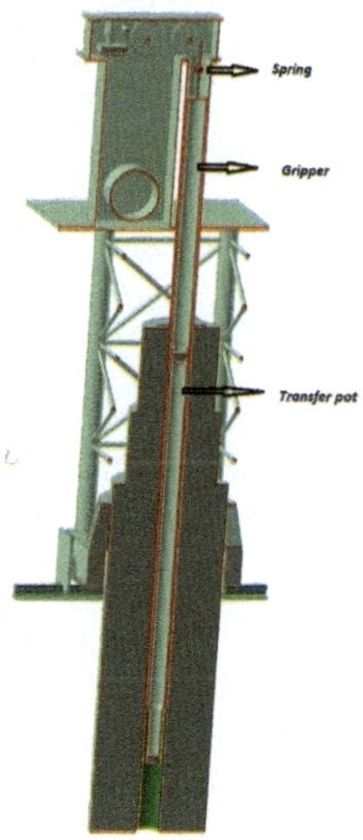

2 Modelling

STF is 3D modelled true to its original dimensions, solid elements are used to model all the components, the upper flask body along with the hoisting mechanism is supported by the support structure, which consists of 4 Fe410WB IS:2062 pipes of Nominal Bore 200 mm and 30 mm thick. The locking hole is connected to rail by rigid elements, and the rail is made as ground in the simulation.

2.1 Simplifications Considered in Modelling

Modelling of STF assembly is considered with the following simplifications:

1. Two lumped masses of 422 kg each are modelled on the upper flask body to represent the mass participation of motors, gear boxes, and other components.
2. Most of the joints except for the four revolute joints between wheel and carriage are modelled as fixed as their functionality is not required and further these joints are removed, and the bodies are directly fixed to reduce the computational time.
3. The two wire ropes that are originally used for raising/ lowering the TP are modelled as a single spring with original stiffness during Tension and negligible stiffness during compression.
4. The gripper fingers are not modelled; instead a spherical joint is given to conservatively estimate the interaction of transfer pot with the inner surface of the flask body. This is justified by the study indicating that during handling operation if any of the three fingers loses contact the TP can swivel about the axis.
5. Spherical joint is used between the modelled rope and pulley instead of modelling the entire rope, as the rope drum is electromagnetically braked and locked, the portion of rope hanging from the rope sheave on pulley has only three rotational degrees of freedom (Fig. 3).

2.2 Properties of Material

1. Fe 10WB IS: 2062 Structural steel.
   Material Density: 7.850e–06 kg/mm^3
   Young's modulus (E): 200 GPa
2. Cast Iron—shielding block.
   Material Density: 7.20e–06 kg/mm
   Young's modulus (E): 110 GPa

Note: As TP and gripper are not modelled in detail and since internals are omitted for brevity, their weight is adjusted to 600 and 550 kg, respectively.

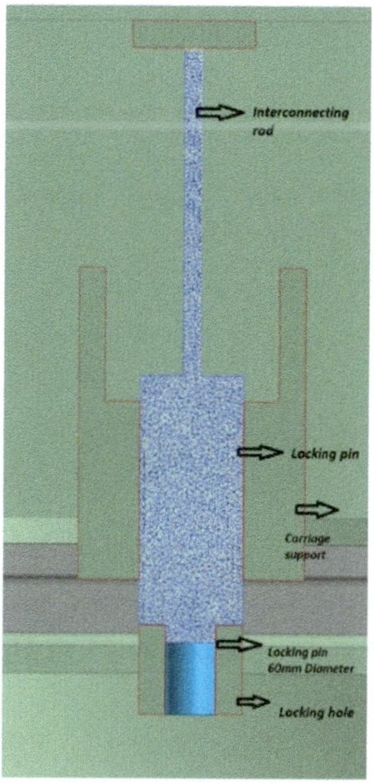

Fig. 3 Locking pin assembly

2.3 Spring Parameters

$$K = AE/L \tag{1}$$

where A is the effective cross-section of Wire rope and is equal to $2 * 0.4 * (\pi/4) * 14^2 = 123.2\ \text{mm}^2$ and L (= 2000 mm) is the length of the wire rope when the load is in maximum lifted condition. E is the Young's modulus of the wire and is equal to 80,000 MPa [2].

$$K = 123.2 * 80000/2000 = 4928\ \text{N/mm}$$

This stiffness value is for wire under tension, there is almost negligible tension during the up motion of TP. So, a spline consisting of varying stiffness during tension and compression is assigned for the spring as shown in Fig. 4.

Damping constant of spring

$$C = 2 \in \sqrt{Km}$$
$$C = 2 \times 0.08 \times \sqrt{(4928 \times 10^3 \times 1150)}$$
$$C = 12.045\ \text{N} - \text{mm/s} \tag{2}$$

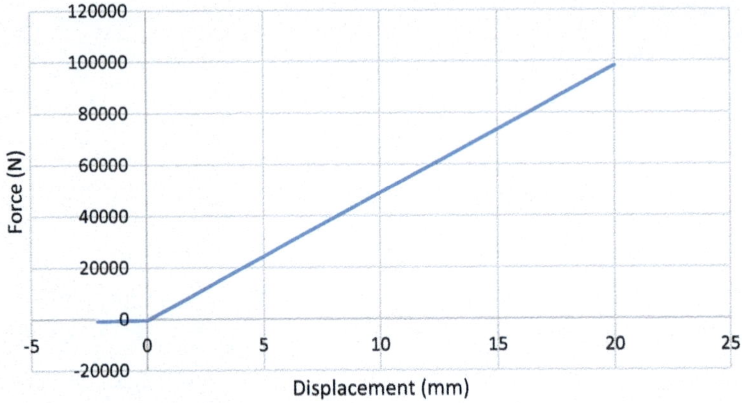

Fig. 4 Stiffness of modelled spring

3 Approach to the Problem

3.1 Boundary Conditions

In the present problem, due to the X, Y(vertical), Z motions of the rail, the STF gets accelerated in respective directions, apart from the frictional force at wheels in X, Z directions, almost entire resisting force is generated in the sleeve of the two locking holes fixed in the ground. The locking pin is pushed into the hole by interconnecting rod from the linear actuator stroke module. Locking pin is a stepped shaft with 60mm diameter pin being locked into the hole, while the 120 mm diameter portion of the shaft is located inside the carriage support structure as shown in Fig. 3. Due to the base excitation, the 120 mm diameter shaft portion comes in contact with the carriage support, which is an annular cylinder with 200-micron clearance from the locking pin. For the locking pin to derive support from the carriage, it must undergo translation and structural bending, so multiflexible body dynamics (MFBD) approach is used. MFBD analysis is generally used for mechanisms having a large number of components. By modelling the system as flexible component, MFBD generates the required force/torque for any required functionality, with the help of FEA or modal reduction method. Recurdyn is the code used for the analysis [3].

3.2 Contact Parameters

Contacts are based on a penalty approach. The contact stiffness is an important parameter in MFBD. The contact parameters have been calculated by using the Hertzian contact theory.

Table 1 Contact parameters

Contact	Stiffness (N/mm)	Damping constant (N-s/mm)
Wheel to rail	2.41e06	243.1
Locking pin to hole	5.93e05	59.39
Locking pin to carriage	8.69e05	86.9
TP to flask body	1.15e06	115.1066

The contact stiffness (K) is given by

$$
\begin{aligned}
\mathrm{K} &= \sqrt{\left(16\mathrm{RE}^2/9\right)} \\
R &= R_1 R_2/(R_1 + R_2) \\
E &= E_1 E_2/(E_1\left(1-\mu_1^2\right) + E_2\left(1-\mu_2^2\right)
\end{aligned}
\tag{3}
$$

R_1, R_2 are radii of curvature of two contacting bodies,
E_1, E_2, μ_1, μ_2 are Young's modulus and Poisson's ratios of corresponding bodies,
And damping coefficient is typically 1/10000 of K.

For Wheel to Rail contact, since the rail is flat, a large value for R_2 has been considered for computing stiffness (Table 1).

These contact parameters are used for respective contacts along with static friction coefficient of 0.24, dynamic friction coefficient of 0.14 for steel to steel contact, and a stiffness exponent of 1.5 for all the contacts [4].

4 Ground Motion Input

The floor response spectra (FRS) have been obtained [1] from dynamic analysis of entire containment building for both S1 (OBE) and S2 (SSE) level of earthquakes. The generated displacement time histories for the response spectrums of Operating Basis Earthquake (OBE) with 2% damping of the horizontal and vertical components are shown in Fig. 5. The time history data generated from this spectrum are applied to the rails, the displacement vs. time graph of rail in X, Y, Z directions are shown in Fig. 6, respectively. Their corresponding accelerations are shown in Fig. 7.

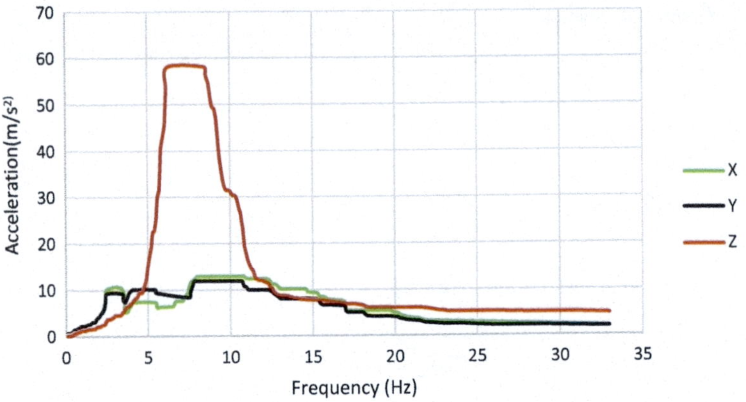

Fig. 5 Response spectrum data

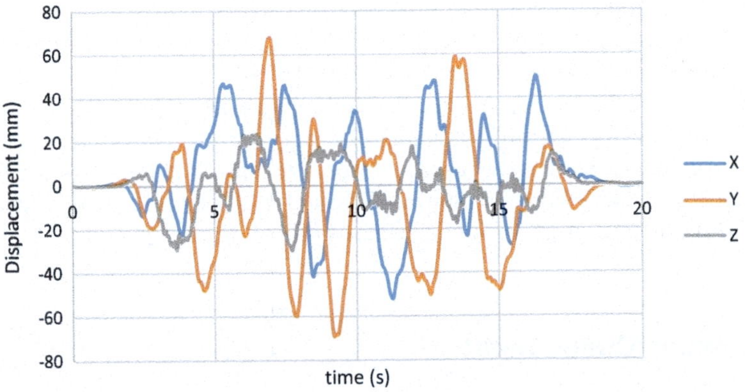

Fig. 6 Displacement time history data

5 Dynamic Force Analysis

5.1 Wheel to Rail Contact Study

Figure 8 is the graph of contact force in Y direction on Wheel 1, the constant force is the steady reaction developed due to self-weight, once the excitation has started the reaction force on these contacts is observed to be varying according to the acceleration of the rails. When the rails are moving upwards (+Y), the STF is pushed upwards and hence the force spikes and when the rails are moving downwards (−Y), the STF behaves like a free-falling body accelerating at g while the rail moves down greater than 1 g acceleration due to added −Y acceleration component of seismic excitation, thereby creating the possibility of wheels leaving the contact with rail.

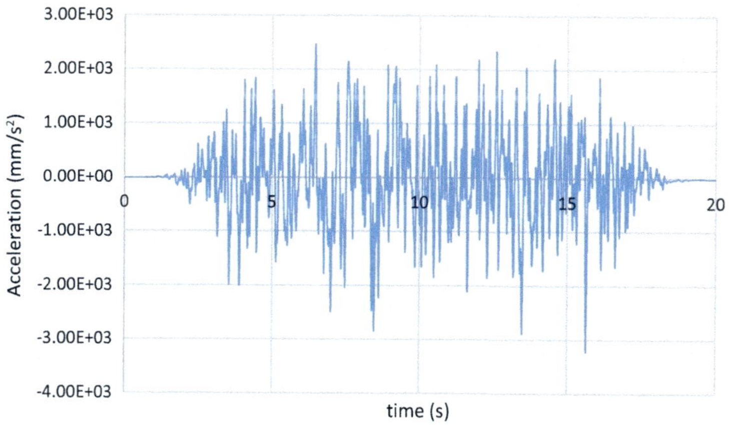

Fig. 7 Acceleration time history in Y direction

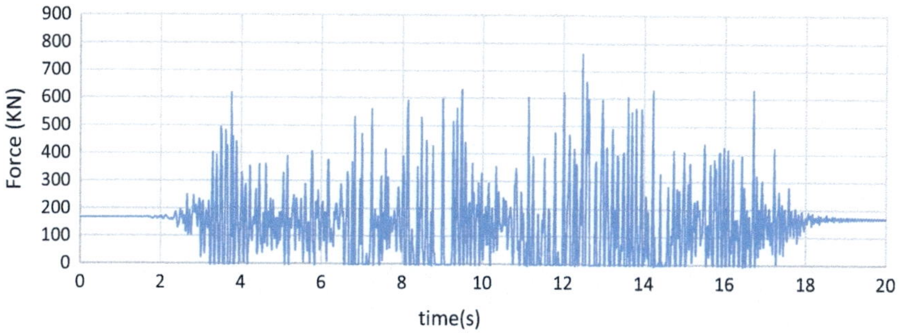

Fig. 8 Wheel 1 (front left) contact force Y direction

As a result, during these intervals, the contact reactions are zero. It is observed that all the wheels don't lose contact at the exact same time (but during the negative acceleration) of rail because of the inertia during upward acceleration and due to the eccentric masses present in the system. The contact force generated is in accordance with the acceleration of rail and dies down at the end of excitation interval. Figures 9, 10 and 11. show the contact reactions for wheel 2, 3, 4, respectively, for 20 s of seismic excitation.

5.2 *Locking Pin Carriage Contact Study*

The lower part of the locking pin is locked inside the locking hole while the 120 mm diameter portion is supported inside the carriage support. Due to seismic excitation, the locking pin displaces and contacts the carriage support at two points. During

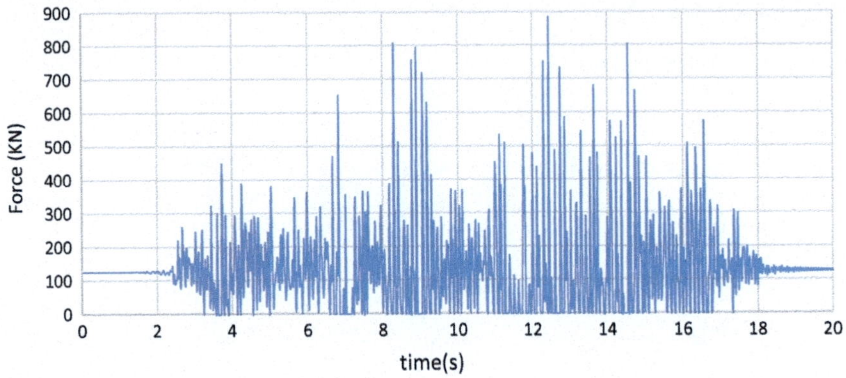

Fig. 9 Wheel 2 (front right) contact force Y direction

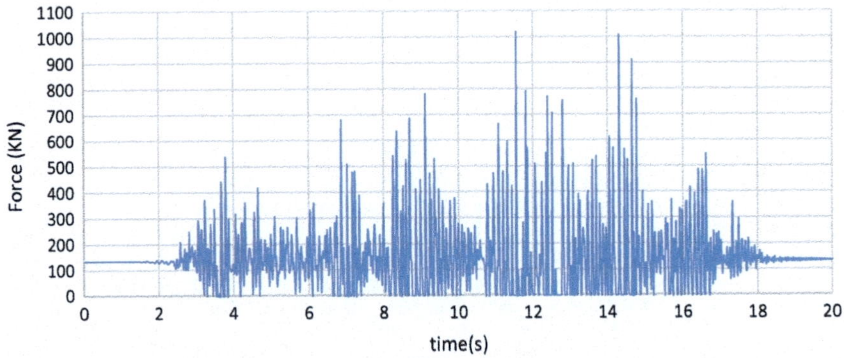

Fig. 10 Wheel 3 (back left) contact force Y direction

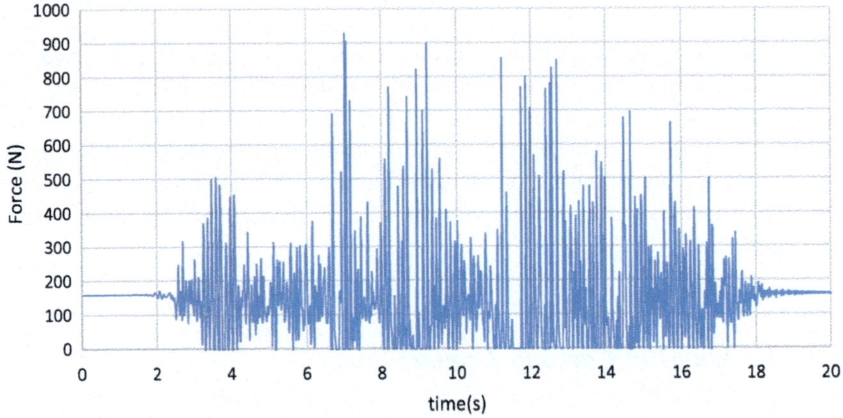

Fig. 11 Wheel 4 (back right) contact force Y direction

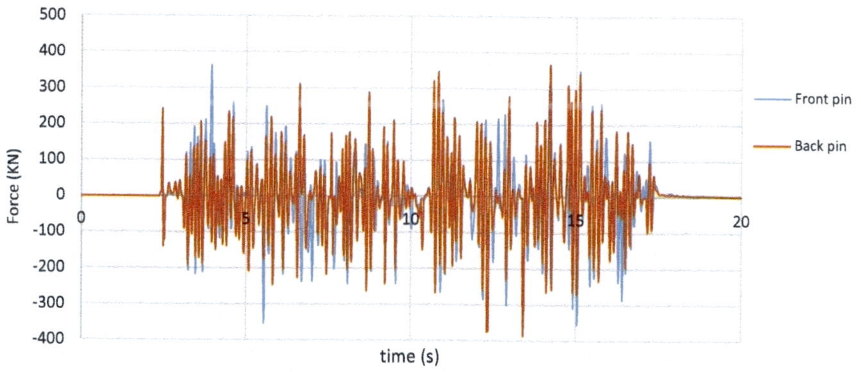

Fig. 12 Locking pin and carriage contact force X direction

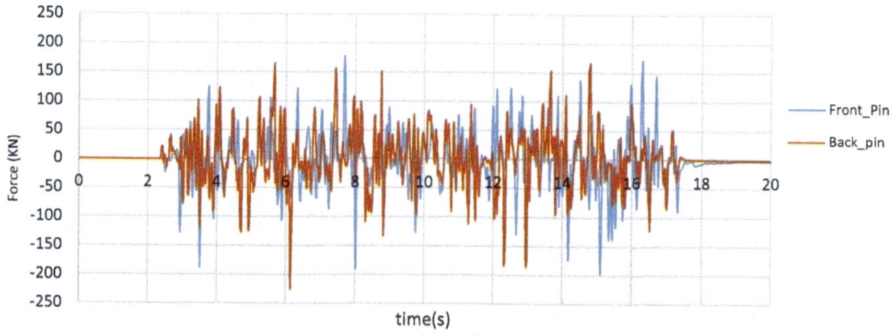

Fig. 13 Locking pin and carriage contact force Z direction

Earthquake motion, the STF is resisted from displacing in the x, z directions by the locking pin reactions and frictional force at the wheels. Figures 12 and 13 show the comparison of contact reactions on Front and Back locking pins in X and Z directions, respectively. Both the locking pins are not experiencing same reactions because of the moment caused due to mass eccentricity about z axis. These forces are used in further analysis for the design verification of the locking pin. At the last 4 s interval, the contact reactions in Z direction are almost same for both front and back pin while the X contact forces are negative for front pin and positive in Back pin indicating the turning moment about positive Y axis. Both forces converge to zero at the end of excitation interval.

5.3 Locking Pin and Locking Sleeve Contact Study

Figures 14 and 15 show the contact forces on the bottom part of the locking pin that is latched into the locking sleeve, theoretically, this portion is completely locked in

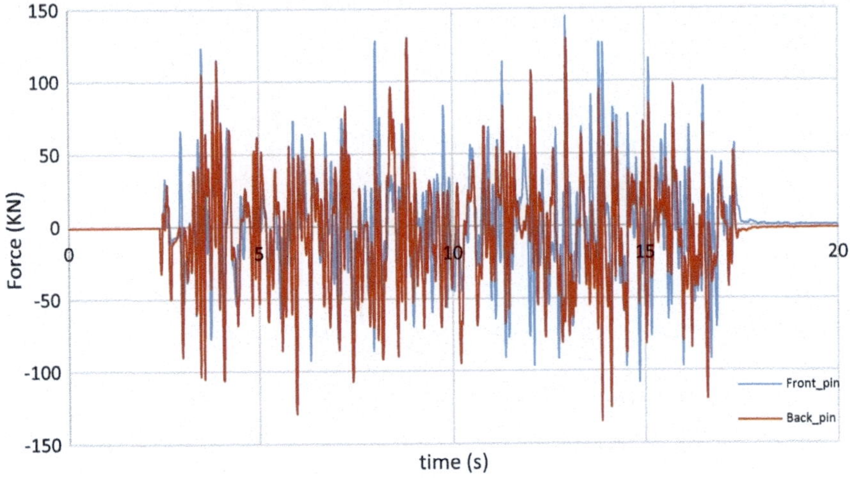

Fig. 14 Locking pin and locking sleeve contact force X direction

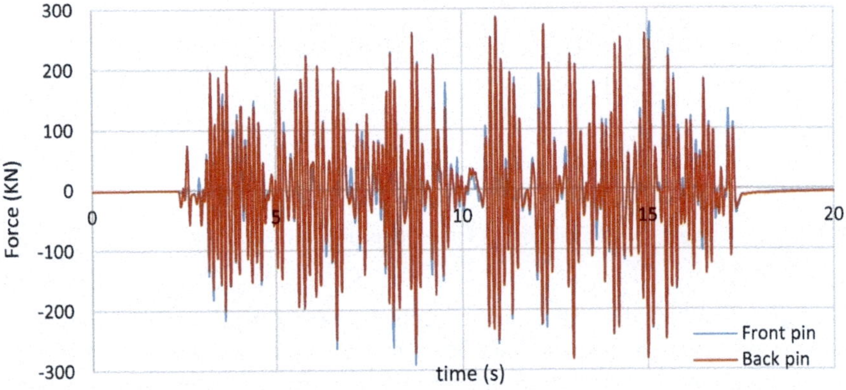

Fig. 15 Locking pin and locking sleeve contact force Z direction

X and Z directions but 200 micron clearance is given between components to ease docking and undocking. The locking sleeve is a part of ground and moves according to the seismic motion. The reactions generated due to the contacts generate high shear stresses in the locking pin. As shown in the graph, initially there is no contact between the pin and sleeve but when the acceleration starts, contact happens and reaction force spikes in the respective directions.

5.4 Study of Transfer Pot Motion

The transfer pot oscillates about the gripper fingers in X, Z direction during the seismic motion and touches the inner surface of the flask body during seismic excitation causing contact forces. Figures 16 and 17 show the contact forces superimposed on the displacement of Transfer pot C.G. There is a clearance of 10mm between the transfer pot and the flask body, even though the CG has not been displaced by 10mm the contact forces are occurring due to the contact happening at the bottom portion of

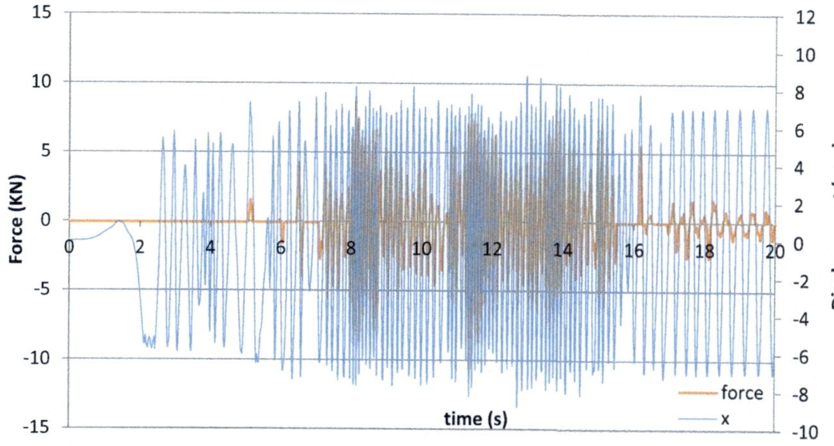

Fig. 16 TP flask body interaction X displacement (TP) imposed on contact force

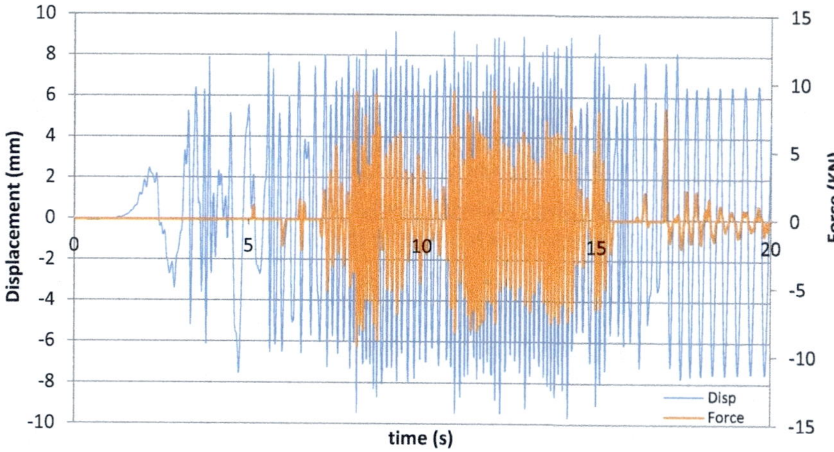

Fig. 17 TP flask body interaction Z displacement (TP) imposed on contact force

the TP at about 3950mm from the gripper support location. As shown in the graphs, the force spikes in respective directions of maximum displacement at that instant.

6 Conclusion

To understand the behaviour of the docked STF during Seismic excitations, a multi-flexible body dynamic analysis is carried out using the MBD software. Contact study of the system indicates that the contact between the rail and wheels is lost in multiple times because of the augmented vertical acceleration due to seismic loading. The stepped portions of the wheel and the seismic arrestor plates have not encountered any contact with the rails during any part of the excitation. The behaviour of gripper and Transfer pot are studied, it is observed that the TP contacts the flask body at about 3950mm from the gripper support location and a maximum impact force of 8 KN is observed. The contact forces on the locking pin from locking hole and carriage structure are estimated. It has been observed that the interconnecting rod of locking pin undergoes structural bending till the contact between the pin and carriage support is established. Forces on locking pin due to the excitation are used for further analysis and design modification of locking mechanism of the STF. The contact reactions during the seismic motion are marginally higher because of the impact occurring at the contacts. Based on the rigid body dynamic behaviour of the STF the functionality effectiveness of the locking pin is verified, and the machine is stable during an Operating Basis Earthquake (OBE 2%) without any rocking and rolling motion (in docked condition).

Appendix

Since the application of multibody dynamic approach to a rapidly changing contact problem, that is seismic behaviour of machine using MBD is an experimental idea and is not widely pursued, the amount of literature available regarding this particular case is minimal. So, a validation study has been conducted to verify the functionality of RecurDyn code specific to the conditions of this study.

A Validation of Multibody Static Analysis

Static equilibrium is achieved when all the bodies are at rest, the gravitational force and loads acting on the system are balanced by the reactions developed by contacts and forces from the joints modelled in the system. In the present case, the contact reactions formed between the wheel and rail balance the entire weight of the machine and the loaded transfer pot. During the contact formation, abnormal contact forces are generated due to impact loads (when the simulation is started, due to gravity, the machine lands on the rails causing impact reactions that are greater than weight of the machine) and wheels bounce back due to the high contact stiffness of structural steel ($F = K^*\delta^n$ (stiffness * depth of penetration$^{\text{Stiffness exponent}}$)). This dynamic behaviour is subjugated eventually by the contact damping provided and the reaction forces stabilize to a more pragmatic value. The static simulation not only validates the constraints, joints and contacts in the system but also acts as the starting point of the dynamic analysis. The static analysis in this problem is indirectly achieved by stabilizing the system for 10 s (so as to bring the system to complete rest without any variation in contact force) as the conventional static analysis is not compatible with motion given to the rails.

The total weight of STF loaded with Transfer pot from FEA = 602,230 N.

Combined reactions from four wheel to rail contacts = 597,916.61.

Percentage error = 0.716%.

Static simulation results are found to be in agreement with the analytical solution with an error of 0.716%. This error is under acceptable limits.

B Validation of Multibody Dynamic Analysis (MBD)

Validation of multibody dynamic model is done by verification of contacts as contact analysis is a nonlinear FEA and is the most intricate part of the MBD analysis. The equations of motion and Newton's second law (F = m.a) are solved in this rigid body analysis. Validation of contacts is important to judge the correctness of the assumed contact parameters and the efficacy of solver methodology while solving a contact problem. Friction parameters, contact damping, contact stiffness are the governing parameters in contact mechanics.

B.1 Validation of Friction

To verify the accuracy of solver in the considering the friction in analysis, a simple block on block model is considered as shown in Fig. 18. The contact between the top block and the bottom block is defined and friction coefficient is set to zero, the bottom block is moved with respect to ground to right and left such that the top block is still supported and contact is never lost. Theoretically, the bottom block should move away while the top block should remain in place due to the absence of contact friction. Simulation resulted in exact depiction of theoretical result, given here is the displacement vs. time graph of two bodies. The bottom body kept on moving while the top body is at rest indicating the absence of friction forces on the body even though the block stays on the top of bottom block due to the support from contact as observed from the contact reaction in Y direction.

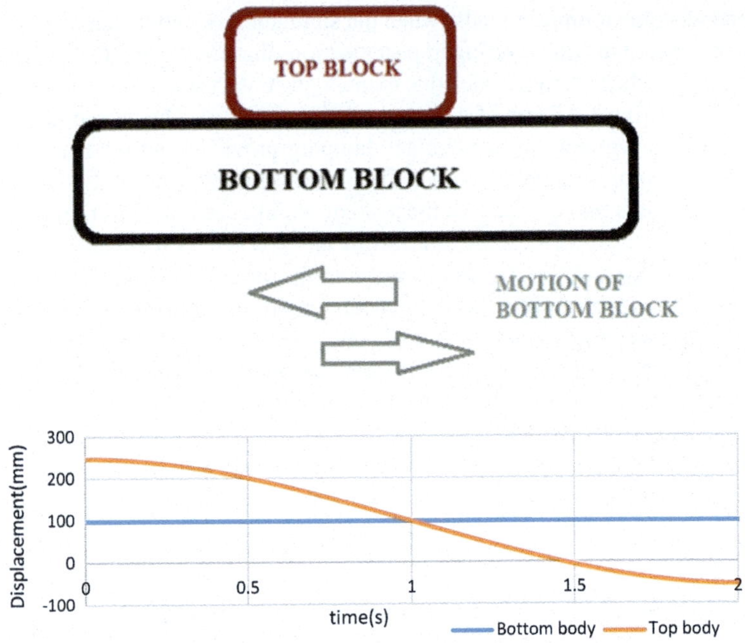

Fig. 18 Arrangement, displacement versus time graph of block on block modal

B.2 Validation of Contact Stiffness and Damping

Contact stiffness and contact damping are the most important parameters for an accurate contact analysis during seismic excitations that are sudden are non-periodic. More stiffness of contact leads to unrealistic bouncing and less stiffness leads to more penetration compared to reality. Similarly, more contact damping leads to attenuation of force from rails to wheels and less damping makes the system bouncy and results in a higher stabilization time than in practical. In the current analysis, the contact stiffness is calculated as per Hertzian contact theory and damping is typically considered as 1/10000 of stiffness value. To validate these input parameters and solver efficiency, data from an experiment are considered:

The modelling parameters are as shown below.

Ball diameter = 13.5 mm.

Plate dimensions are 100 × 100 × 100 mm.

Both plate and ball are made of steel.

Young's modulus = 2.1 e05 MPa.

Poisson's ratio = 0.3, Density 7800 kg/m^3

Drop height h_1 = 500 mm.

According to [5], the first bounce of the steel ball (h_2) estimated from several drop test experiments is between 160 and 170 mm. An average value of 165 mm is assumed for error calculation from the simulation result.

Coefficient of restitution (k) according to experimental data is

$$k = \sqrt{(h_2/h_1)}$$

$$k = \sqrt{\left(\frac{165}{500}\right)}$$

$$k = 0.57445$$

From the simulation, the bounce height is estimated to be 173 mm, so the coefficient of restitution based on the assumed contact parameters is 0.5882. The location of C.G of ball Vs time after dropping is shown in Fig. 19.

Error in coefficient of restitution between experimental [5] and simulation is

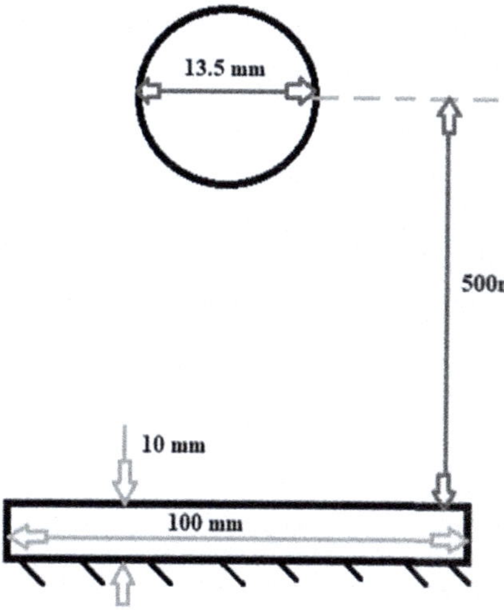

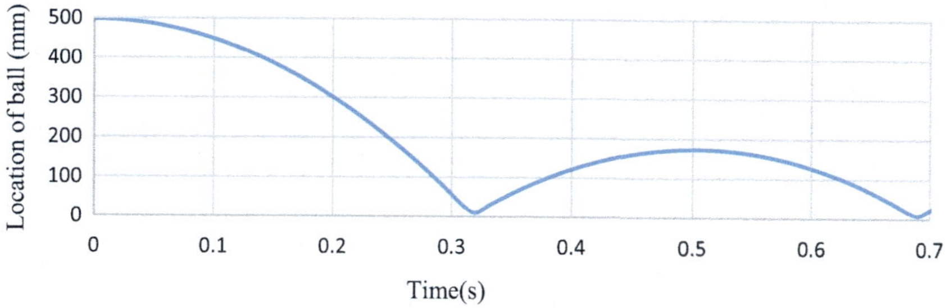

Fig. 19 Location of C.G of steel ball versus time after dropping

$$\begin{aligned} Error\% &= \frac{0.5882 - 0.57445}{0.57445} \\ Error &= 2.4\% \end{aligned}$$

This error is within acceptable limits, and the contact parameters used in the analysis are properly validated.

C Validation of Flexible MBD Study

In the STF model used in the analysis, the locking pins are modelled as flexible elements as they have to deform little to extract support from the carriage (clearance is 1 mm). To justify that the solver is analysing the flexible elements properly, the Locking pin is modelled true to its dimensions and made into flexible element. It is in the form of a stepped cantilever beam supported as 25 mm diameter end. The deflection caused by the free end due to self-weight is verified by comparing it to mathematical (as shown in Fig. 20) and FEA analysis result. Material properties of locking pin are used for the calculation.

Analytical Result

By conjugate method, deflection of free end in original beam is equal to the bending moment at fixed end on an equivalent conjugate beam with loading diagram as M/EI diagram of original beam.

Calculations through conjugate beam method give deflection at free end as 3.217 mm.

MBD Flexible Analysis

Deformation at the end of cantilever type locking pin through MBD analysis is 3.1040 mm.

$$Error = \frac{3.217 - 3.1040}{3.217} = 6.3\%$$

Same geometry of stepped shaft is subjected to deformation because of self-weight using flex body module in RecurDyn as shown in Fig. 21. The error in the deformation at free end is approximately 6.3% and is within the acceptable limit and thus the flexible body of STF is properly solved in the MBD analysis.

The above analysis validates the parameters and the method of approach used in the multibody dynamic study of subassembly transfer flask under seismic excitation.

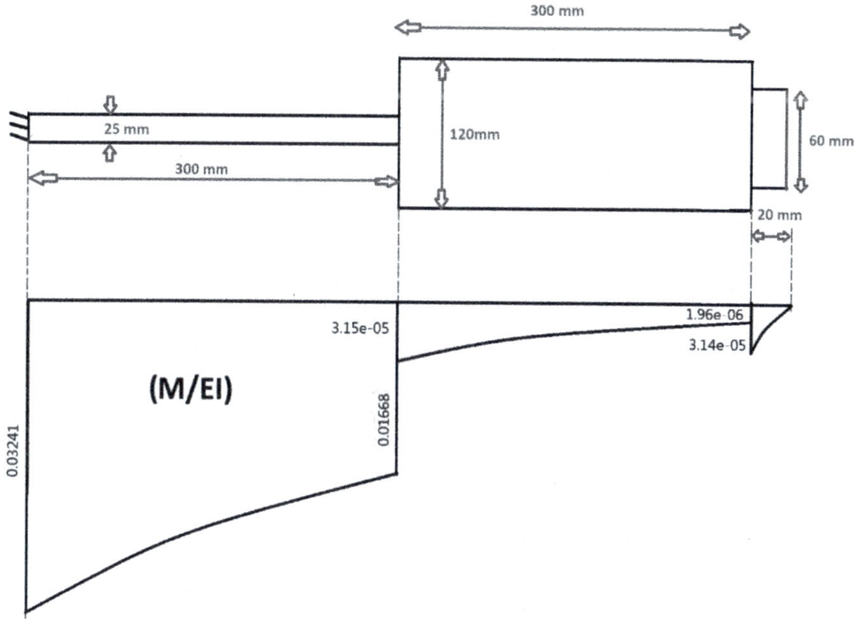

Fig. 20 (M/EI) diagram of the stepped beam

Fig. 21 Deflection of beam under MBD analysis

References

1. Sajish SD, Ramakrishna V, Chellapandi P, Chetal SC (2003) Generation of floor response spectra for PFBR RCB. In: National seminar on seismic analysis of nuclear power plants, SERC, Chennai, 21–22 Feb 2003
2. Kumar S. Design of single failure proof hoist for subassembly transfer flask. IGCAR internal report
3. Recurdyn documentation—Recurdyn user's manual
4. Seismic design criteria for structures, systems and components in nuclear facilities. ASCE 43-05
5. Liu X, Chen W, Shi H (2022) Improvement of contact force calculation model considering influence of yield strength on coefficient of restitution

A Finite Element Analysis Study on the Effect of Tool Stiffness in Incremental Forming Process

Eldho Paul, Hariharan Krishnaswamy, Riby Abraham Boby, and Sahil Bharti

Abstract Recent developments in the area of single-point incremental forming process deal with the study of forming accuracy. In general, the forming tool is assumed to be rigid, contrary to what exists in real experiments. This paper presents the simulation results of the effect of tool stiffness in the single-point sheet incremental forming process. A finite element method-based analysis was adopted for this study. The analysis is carried out using tools having different compliance. The nodal displacement of the deformed profile of the truncated cone was measured and compared with the results from the numerical trials with a rigid and deformable tool. The study indicates that using a deformable tool results in considerable geometric errors in the formed component.

Keywords Incremental sheet forming · Finite element method · Tool stiffness · Deformation

1 Introduction

Incremental sheet metal forming (ISF) is a dieless metalworking process that has proven to be a promising technique for manufacturing components with complex geometries, which were previously difficult to achieve with traditional metal forming processes. In ISF, the desired shape is obtained by forcing the sheet metal with

E. Paul (✉) · H. Krishnaswamy (✉) · S. Bharti
Mechanical Engineering Department, Indian Institute of Technology, Madras, Chennai 600036, Tamilnadu, India
e-mail: eldhopaul@mace.ac.in

H. Krishnaswamy
e-mail: hariharan@iitm.ac.in

S. Bharti
e-mail: me17d017@smail.iitm.ac.in

R. A. Boby
Mechanical Engineering Department, Institute of Technology Jodhpur, Jodhpur 340037, Rajasthan, India

S. K. Saha and K. Rama Krishna (eds.), *Recent Developments in Multibody Dynamics*, Lecture Notes in Mechanical Engineering,
https://doi.org/10.1007/978-981-97-7525-5_12

Fig. 1 ISF process on CNC machining centre

a simple ball-end tool that follows a predetermined path [1]. This process can be controlled by Computer Numerical Control (CNC) machines or robots, which guide the tool along the specified trajectory on the metal surface. The forming tool induces local deformation, causing the sheet to plastically deform and take on the intended shape. A CNC machining centre with ISF tooling is shown in Fig. 1. CNC-based ISF is favoured for its accessibility and ease of development and use. However, CNC machines are limited in workspace and flexibility for creating complex parts due to their lower degrees of freedom. Alternatively, robots provide a higher degree of freedom and a larger workspace for ISF. Although robots provide enhanced flexibility, their overall stiffness is lower than CNC machines [2].

Parametric studies in ISF often focus on factors like sheet thickness, tool diameter, feed rate and tool path methodology [3]. Machine and tool compliance is often overlooked, assuming rigid behaviour throughout the process [4]. The overall tooling in the forming process includes various components, such as the programmable machine, tool holder and the forming tooltip. Neglecting the individual compliance of tooling can lead to incomplete assessments of process dynamics. Elastic deflection of machine tools during forming can affect the contact force between the tool and workpiece, resulting in the inaccurate prediction of sheet thinning and leading to early fracture [5]. The effective tool-sheet contact tends to differ from the programmed tool path due to the tool deflection under forming loads. This influences the geometric accuracy of the formed component. CNC machines are generally considered 15 times more rigid than the forming tool, as shown by an experimental study [6].

Therefore, the study presented here examines the impact of tool compliance on the geometric deviation of ISF-formed components. Typically, an experimental investigation is preferred using commonly available measurement instruments such as coordinate measuring machines, laser trackers, and monocular cameras, which ensure a higher degree of accuracy. However, it is practically challenging to record real-time tool deflection during ISF due to the dynamic nature of the process and the limited workspace of CNC machine tools. Additionally, using instruments such as laser trackers presents difficulties in maintaining an uninterrupted line of sight during forming operations. Therefore, FEA-based analysis is adopted here. The FEA-based study is considered an effective tool for analysing the metal forming process [7]. The machine or tool deflection results obtained from the FEA simulations can be directly used for either the tool path compensation or real-time corrections in the tool path trajectory to reduce the profile inaccuracies during forming.

A FE analysis of the ISF process was conducted using rigid and deformable tools. A truncated cone was developed. The geometric profiles of the deformed part using rigid and deformable tools were compared. Here, the tool is modelled as a deformable shell structure. The nodal displacement of the deformed profile of the truncated cone is compared for each case. The role of the tool stiffness on the forming prediction in the incremental forming process is analysed.

2 Finite Element Modelling

A production process of a truncated cone using a Single Point Incremental Forming (SPIF) process was studied by numerical simulations using a commercial software code, Abaqus/Explicit. An elasto-plastic finite element model is set up to simulate the SPIF process. Dynamic explicit simulation with step time 10 and mass scaling of 0.0001 is used for the simulation. A geometric model of the sheet and tool is assembled in the software's workspace, as shown in Fig. 2. A blank of 150 mm × 150 mm is modelled as a three-dimensional deformable and discretized by S4R four-noded doubly curved shell element of global size of 1 mm with hourglass control. Around 22800 quadrilateral structured elements were generated after meshing of the deformable sheet. The material used for this study is Aluminium 1050 (AA1050) sheet of 1.2 mm thickness. The chemical composition of AA 1050 is shown in Table 1 [8]. The flow stress curve of the material is obtained from the combined hardening law, and material parameters are adopted from the study [8].

An 8 mm diameter ball end tool is used in this simulation. Initially, the tool is modelled as analytical rigid for one set of simulations. In the rest of the numerical trials, the tool is modelled as deformable body. The deformable tool is discretized by S4R four-noded doubly curved shell element of global size of 1 mm. Interaction between tool and sheet is defined using a penalty-based friction model with a coefficient of friction of 0.1. Hard contact is used to define the normal contact response.

Table 1 Chemical composition of Aluminium (AA1050) sheet

Element	Al	Si	Fe	Cu	Mn	Mg	Others
Weight(%)	99.8	0.0926	0.0639	0.0203	0.00722	0.00886	0.0433

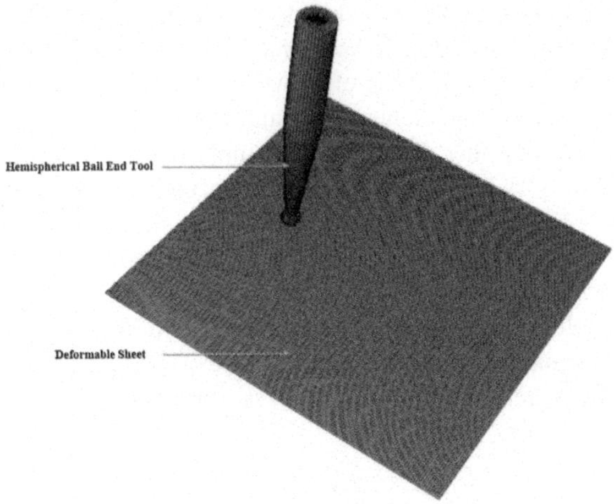

Fig. 2 Finite element model of the SPIF process

Deformable tool material elastic properties were modelled by varying Young's modulus and keeping the other parameters constant as that of high-speed steel-M2. The effect of compliance obtained from each analysis is compared.

The truncated cone tool path for the simulation is created using commercially available Fusion 360 modelling software. The spatial coordinates of the tool path were extracted from the G-code and these spatial coordinates and time were provided as an input to define the boundary conditions amplitude of the tool in Abaqus simulation.

3 Results

The stiffness of the tool was systematically varied by altering Young's modulus. The rest of the parameters, such as the constitutive relation of the workpiece material, interface friction and other boundary conditions, remained invariant between the numerical trials. These cases were simulated numerically where the tool young's modulus was equivalent to 150 GPa, 200 GPa and completely rigid. Figure 3a, b, respectively, shows the deformed profile and displacement magnitude of the sheet in the SPIF process with a rigid tool and deformable tool with 200 GPa Young's modulus. Figure 4 shows an overlay plot of the deformed profile of the sheet with a rigid and 200 GPa tool. The colour code shows the deviation in the deformed profile with rigid and 200 GPa tool, where red colour indicates the maximum deviation area, green represents the area with average and blue shows the least deviation area. Variation in the forming depth is larger in the wall area than in the rest of the sheet. The overlay profile shows a regular pattern of deviation throughout the sheet except in some regions. It can be inferred as an effect of higher localized tool deflection.

Table 2 shows the comparative results obtained from the simulation study of two experiments with a 200 GPa deformable tool and a rigid tool. Comparing the linear

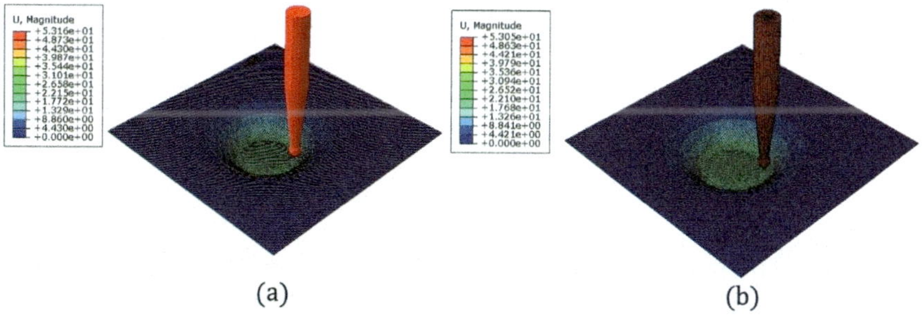

Fig. 3 Deformation profile of truncated cone in SPIF process with a **a** rigid tool and **b** 200 GPa tool

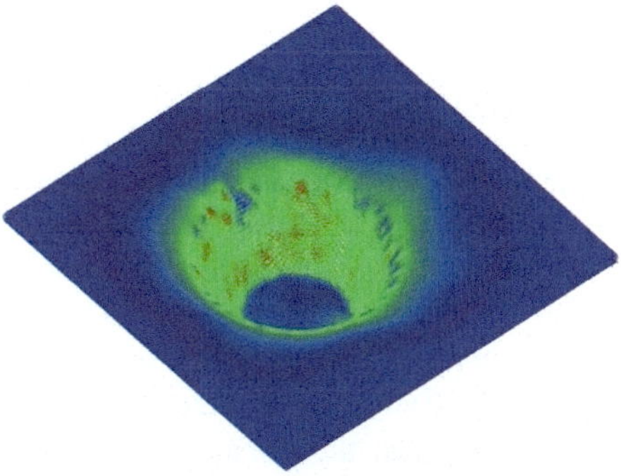

Fig. 4 Overlay plot of the truncated cone in SPIF process with rigid and deformable tool

Table 2 Comparison of profile displacement with 200 GPa deformable tool and rigid tool

Direction	Maximum deviation (mm)	Mean deviation (mm)	Standard deviation (mm)
X-direction	0.437	0.002	0.065
Y-direction	1.157	0.174	0.204
Z-direction	0.396	0.002	0.066
Nodal displacement	1.188	0.174	0.205

displacement, a maximum nodal displacement deviation of 1.157 mm was found in the Y direction, which is parallel to the tool's axis. This displacement deviation will have a significant effect on the forming accuracy for a given 20 mm tool displacement boundary condition in the Y direction. The mean deviation of nodal displacement is 0.174 mm with a standard deviation of 0.205 mm, which shows how tool stiffness has an impact on the accuracy of the SPIF deformed profile.

In this work, the tool is assumed to be deformable, while the rest of the machine, including the tool holder, is assumed to be rigid. In other words, the net effect of the entire chain of machine elements stiffness is lumped into the tool stiffness. For the FEA simulation, the Young's modulus of the tool is assumed as 200 GPa, which corresponds to the tool material. The result observed from the simulation shows a mean nodal deviation of 0.174 mm. The effective Young's modulus of the lumped model needs to be substituted to understand the lumped effect. Effective Young's modulus of the machine element and single point tool arranged in series is given as,

$$1/E_{eff} = 1/E_{tool} + 1/E_m \tag{1}$$

where E_{eff} is the effective stiffness, E_{tool} is the stiffness of the slender tool, and E_m is the stiffness of the machine and tool holder system. Based on the literature [6], a rough estimate of the effective modulus of the machine and tool system E_m is at least 15 times that of the tool. Substituting this in Eq. 1 gives an effective stiffness of 187.9 GPa for this arrangement.

The effective role of the entire chain of machine elements, including frame, holder and tool, on SPIF can be understood by substituting lumped stiffness 187.9 GPa in the FE analysis. The stiffness of the machine and tool holder system is 15 times of the tool, which is a rough estimate. For shorter and thicker tools, it will be further reduced. As a general case, a conservative lower estimate of 150 GPa would yield an approximate ratio of E_m/E_{tool} to 3 times. A simulation study was conducted to find an extreme case of the role of the entire tool elements using 150 GPa and a mean nodal deviation of 0.180 mm was observed for the same. Figure 5 compares the deformation by 200 GPa and 150 GPa. It is noted that even when assuming the modulus of the tool to be 150 GPa, the deviation in the deflection of the final component is negligible when considering the tool and lumped stiffness. Therefore, the assumption regarding the stiffness of the deformable tool, rigid holder and machine are reasonable for the present work.

The deformation path of the sheet is analysed, as shown in Fig. 6. The graph clearly shows the variation in the forming depth throughout the profile while using

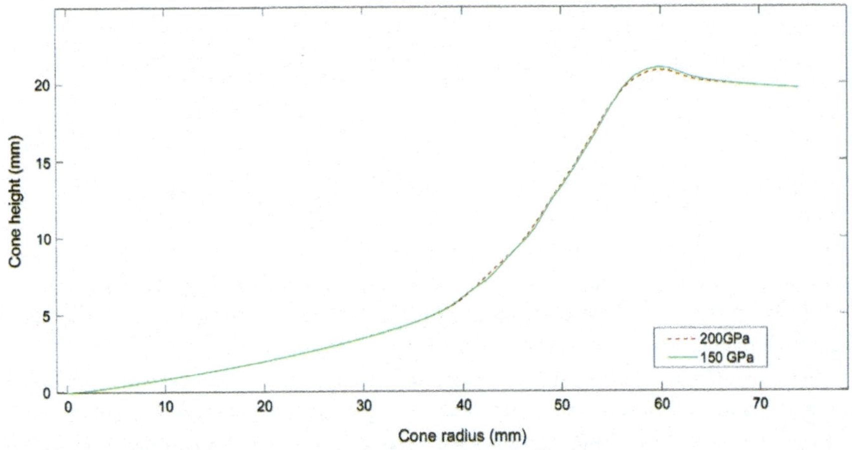

Fig. 5 Comparison between the deformation by 200 GPa and 150 GPa

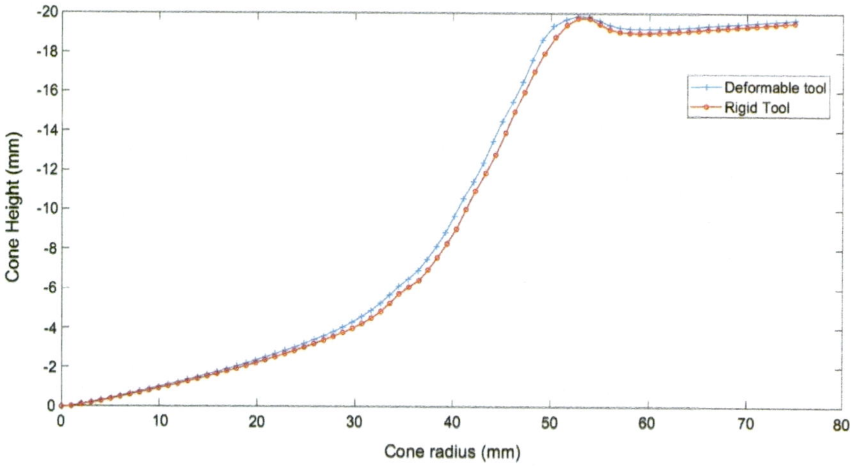

Fig. 6 Deformed cone inner surface profile

the deformable and rigid tool. The deformation in the wall region is comparatively higher than the bottom region of the cone profile. This shows the importance of compliance in sheet incremental forming. During the simulation, the deformation pattern of the single node with respect to time increment is also analysed, as shown in Fig. 7. The nodal displacement for both rigid and deformable tools shows a similar contour during the initial time increment of simulation. Later, the deviation in the nodal displacement is significant as step time increases, but the rate of deformation is same since the node 14602 is located in the wall region of the profile.

Stress distribution in the deformed region of the plate in the conducted experiments shows a variation along each nodal point, but this variation does not exhibit a similar pattern to the displacement profile since springback and other plastic deformation effects contribute to the nodal stress value.

4 Conclusion

This study focused on investigating the effect of tool stiffness on sheet incremental forming. FEA-based simulation experiments were conducted to study the influence of tool stiffness on the accuracy of the deformed profile. In this simulation, all the simulation parameters were kept the same except tool stiffness. For a particular model, the simulation is done with a rigid tool and a deformable tool and the results are compared. This similar simulation environment gave results with good accuracy for the comparative study of tool stiffness. A maximum deviation of 5.78% was found in the deformed profile of 200 GPa compared to rigid tool simulation. This indicates that the deviation is significant for deformable tools, which provide a future scope

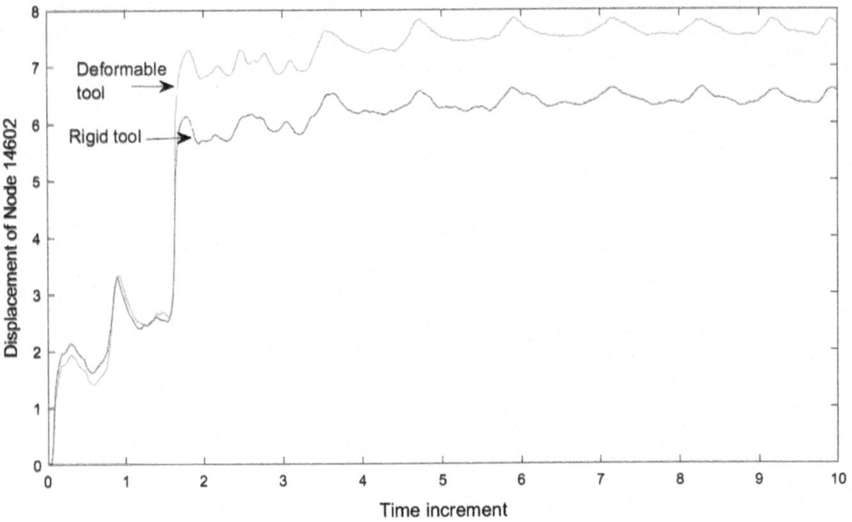

Fig. 7 Deformation history of node 14602 during the forming process

for analysing slender and long tools and robot-like machines with high compliance, which can exhibit higher deviation. The comparative study of tool stiffness 200 GPa and equivalent stiffness 150 GPa gave an insight that tool stiffness is the significant one among the entire chain of machine elements.

Acknowledgements This work was supported by DST via project no. DST/INT/RUS/RSF/P-50/2021 titled "Enhancing the accuracy of roboforming through prediction and compensation of elastic behaviour using Artificial Intelligence techniques". Part of this work was supported by the RSF (project number 22-41-02006).

References

1. Leszak E (1967) Apparatus and process for incremental dieless forming. Patent US3342051A1
2. Klimchik A, Bondarenko D, Pashkevich A, Briot S, Furet B (2014) Compliance error compensation in robotic-based milling. In: Ferrier JL, Bernard A, Gusikhin O, Madani K (eds) Informatics in control, automation and robotics. Lecture Notes in Electrical Engineering 283
3. Jeswiet J, Micari F, Hirt G, Bramley A, Duflou J, Allwood J (2005) Asymmetric single point incremental forming of sheet metal. CIRP Ann 54:88–114
4. Ren H, Moser N, Zhang Z, Ndip-Agbor E, Smith J, Ehmann KF, Cao J (2015) Effects of tool positions in accumulated double-sided incremental forming on part geometry. ASME J Manuf Sci Eng 137(5)
5. Moser N, Leem D, Ehmann K, Cao J (2021) A high-fidelity simulation of double-sided incremental forming: Improving the accuracy by incorporating the effects of machine compliance. J Mater Process Technol 295

6. Salgado MA, López de Lacalle LN, Lamikiz A, Muñoa J, Sánchez JA (2005) Evaluation of the stiffness chain on the deflection of end-mills under cutting forces. Int J Mach Tools Manuf 45:727–739
7. Dejardin S, Thibaud S, Gelin JC (2008) Finite element analysis and experimental investigations for improving precision in single point incremental sheet forming process. Int J Mater Form 1(Suppl 1):121–124
8. Bharti S, Gupta A, Krishnaswamy H, Panigrahi SK, Lee MG (2022) Evaluation of uncoupled ductile damage models for fracture prediction in incremental sheet metal forming. CIRP J Manuf Sci Technol 37:499–517

Mobile Haptic Device for Large Virtual Environments

P. Abinaya, K. S. Sasivarnan, Prasanna Kumar Routray, and M. Manivannan

Abstract Virtual Environments (VE) need large workspaces. On the other hand, haptic devices are mostly grounded and typically have small workspaces compared to the VE. The main objective of this work is to extend the smaller workspace of haptic devices for Virtual Reality (VR) applications without affecting the haptic perception of the user. A position drift-based robotic approach has been proposed in which a redundant kinematic Degree of freedom (DOF) is added to the haptic device by moving its base using a mobile robot to expand the workspace of the haptic interface point (HIP) in the VE. In order to validate the proposed approach, a psychophysical study measuring stiffness Just Noticeable Difference (JND) has been conducted to compare the user perception before (Stationary Robot) and after the proposed approach (Moving Robot). Two experiments were conducted separately, one for a single axis (Experiment I) and another for a two-axis (Experiment II). The %JND for Experiment I is 21.08% and 24.59% for stationary and moving haptic devices. The %JND for Experiment II is 20.55% and 26.56% for stationary and moving haptic devices. One-way ANOVA was performed to test whether the difference in the %JND between the two cases is insignificant. With the insignificant %JND difference ($p = 0.32$ for Experiment I, $p = 0.34$ for Experiment II), the proposed method could be considered as not affecting the user haptic perception while expanding the haptic workspace for the interactions in VE. The stability of the haptic device has been analyzed, including robot dynamics. The current study implements up to two axes of workspace expansions for the 3D systems Touch haptic device; however, the algorithm is extensible to three axes and any grounded haptic devices.

Keywords Mobile haptics · Virtual reality · Workspace · Psychophysics · Stability

P. Abinaya (✉) · K. S. Sasivarnan · P. K. Routray · M. Manivannan
Department of Applied Mechanics & Biomedical Engineering, Indian Institute of Technology Madras, Chennai, Tamil Nadu, India
e-mail: mani@iitm.ac.in

S. K. Saha and K. Rama Krishna (eds.), *Recent Developments in Multibody Dynamics*, Lecture Notes in Mechanical Engineering,
https://doi.org/10.1007/978-981-97-7525-5_13

1 Introduction

One of the main applications of the haptic device is improving immersion in virtual environments (VE) to enable users to touch and manipulate objects in VE. Haptic displays are generally manipulators that provide an interface to the user through its end effector [1, 2]. Current haptic feedback devices are impedance or admittance systems that are mostly grounded [3–5]. Examples of devices include the force feedback mouse, joystick, and Pen-based haptic devices. These devices have limited workspaces and are unsuitable for human-scale interactions [6, 7] in VE. There are also haptic feedback gloves, wearable fingertip haptic devices [8], exoskeletal devices, and haptic suits, which come under wearable haptic technologies. However, these devices have limitations such as high inertia, less stiffness rendering, and cannot render an accurate gravity force. Large-scale devices such as the Scaleable SPIDAR, HapticMaster, WearableMaster [9], Virtuose6D, CyberForce, Inca 6D, and VirtuaPower are available [7, 10]. However, they have limitations in terms of design complexity and reduction in the device performance. Here, we have developed a novel approach for creating large workspace haptic devices by expanding the workspace of small grounded haptic devices with a simple design without affecting the existing characteristics of the device.

In VR applications, an avatar or virtual tool is mapped with an end effector position. As the user interacts with the VE, the corresponding reaction force is calculated using force rendering algorithms such as the finger proxy algorithm and then rendered to the user through the actuator control of an end effector [11]. Most VR environments require the user to move in a larger workspace; however, haptic devices are grounded and have a smaller restricted workspace. This restricted workspace makes it challenging to use with VR applications in which the VR scene contains a larger workspace than the haptic device where objects are positioned far apart [12]. One of the examples of this is that of VR and haptic-based training in emergency medical procedures. The VR scene of such procedures would contain the entire operation theatre with many devices. Therefore this study focuses on solving the workspace limitation problem of the haptic feedback devices using robotics and position-based workspace drifting method.

A few methods based on the fixed table approach have been reported in the literature [6]. A bidirectional motion table [13] used for expanding the workspace of the haptic device is one of the methods. Although this is an excellent method to increase the workspace, since the table is fixed, haptic feedback is restricted to the size of the table. Increasing the workspace for the haptic device using a redundant sliding axis [13] is another method. Redundant Robots may either be designed or attained by adding large range axes to an existing haptic device. This technique allows for a large range of actions the user could carry out in a virtual assembly process or projection-based on large VR systems such as cave automatic virtual environments (CAVE) [14]. Its limitations are its cost and inability to shift it from the room as it is fixed. Similar to these methods, a few more designs are based on the actuation approach [1]. Most of them have limitations regarding design portability and range of movement. Using a mobile robot could solve this problem as the robot may move along with the user for long distances.

Changing the workspace of the virtual haptic tool through control algorithms is another technique to increase the haptic interaction workspace. Scaling control [15] is one of the control methods in which the workspace scale factor of the virtual haptic tool is increased, but this affects the resolution of movements in the VR environment. The indexing method is another simple technique in which repositioning the haptic device is used to reset and recreate the necessary workspace for the virtual tool. However, this requires an intermediate disconnection between the haptic device and the virtual haptic tool, affecting continuous haptic perception. In the Ballistic control method [15], the workspace scale between the haptic device and the virtual tool is varied based on the velocity of movement of the end effector. This method has an offset problem in which there is a shift in position between the end effector and the virtual haptic tool. In the Rate control method [16], the velocity of the virtual haptic tool is varied proportional to the end effector displacement, which causes the motion of the virtual tool even though the end effector is at rest. In positioning with a modified rate control method, an additional switch is required to control the workspace expansion algorithm, which causes an interruption in the haptic perception [7, 17, 18]. Another method called workspace drift control [12] is implemented by considering the perceptual nature of the user in which the workspace of the virtual haptic point is drifted in a way the user cannot perceive. Since this method requires an object interaction during the workspace drift, a VR scene that involves the interaction of many objects located far away cannot be achieved. Hence, a control algorithm that changes the virtual tool's workspace without interrupting the haptic perception and user motion in VR is required.

There are three robotic-based haptic workspace expansion techniques that have been reported in the literature which are similar to the proposed method. A new haptic system was developed that allows the simulation of multi-contact grasping tasks in VE, through the use of two standard Haptic devices grounded on a mobile robot called Nomad XR4000 [19, 20]. They developed a smoothing technique that improves the quality of the haptic rendering using a Mobile Haptic Interface (MHI) [21]. Even though this method is effective, this device is only limited to the Nomad XR4000 robot and the smoothing technique used introduces an error in the Moblie Platform position estimation. Another robotic-based haptic interface [22], which enables a Human Operator (HO) to freely walk around and haptically interact with extended VE. However, this has limitations in terms of portability and design complexity. Another mobile haptic interface called POSTECH MHI (pMHI) [23], has been introduced with an emphasis on its system improvements. This device has not yet been tested in VE which requires higher accuracy Haptic Interface Point (HIP) tracking and is also not suitable for dynamic VE. All these existing devices did not study the device performance for user perception.

Recently, perceptual studies have been emphasized in VR and haptic-based applications. Generally, the simulated visual and force feedback through the VR and haptic systems has been quantitatively validated by measuring the end output characteristics of the system, such as frame rate, haptic loop rate, maximum renderable force, z-width, and damping [1, 10]. However, these validations do not consider the human in the system loop, who is the ultimate user of the system. Therefore, analyses that include humans could provide more detail on system usability. Human perception can be quantified and analyzed through psychophysical measures. One of the basic psychophysical measures is Just Noticeable Difference (JND), defined as the sensory system's sensitivity. Recently, many VR and haptic simulations have been validated by measuring JND [24]. The workspace increase approach proposed in this work also has been analyzed using psychophysical experiments measuring stiffness JND.

The haptic device and stability of the same are coupled and analyzed in the past [25–27]. In this study, we have tried to understand the effect of a modified workspace on the stability of a haptic device. The region of passivity, as suggested in [27], is a result of the non-linear nature of haptic devices. Authors [27] have tried to analyze the behavior of haptic devices under the passivity criterion. Some devices operate stably even in the passive region, where they cannot dissipate energy fully. In [25], a haptic device is studied for the stability boundary to see the stable, unstable, and passive regions for a particular device configuration. Authors in their work [26] addressed the effects of discretization, delay, and Coulomb effects. With the new configuration, the change in device parameters will lead to a change in the stability of the device which is discussed in Sect. 2.6.

Most of the drawbacks and limitations in existing methods to increase the workspace of grounded haptic devices are addressed in our work. The main contributions of the current work are given as follows:

- We propose a simple system design to increase the workspace of small grounded haptic devices without affecting their dynamics. The design is cost-effective, portable, and scalable to three axes.
- A mobile robot is designed to increase the workspace of the haptic device on a larger scale. Since it is not fixed and independent of the haptic device, the design of the robot is easily extendable to different specifications.
- A position control algorithm is proposed to increase the virtual workspace of the haptic tool in the VE. The algorithm is independent of the device, extendable to three axes, and does not interrupt haptic perception.
- A user perceptual study is conducted to analyze the implemented approach of workspace expansion instead of using only the system parameters. This gives a new approach to analyzing the implementation when visual and haptic simulations are involved.
- An essential system parameter to be considered during the design of the robot to ensure the haptic device stability is also discussed.

In summary, the objective of the work is to increase the workspace of haptic devices for VR applications without affecting the haptic perception of the user. For this, a robotics approach has been proposed in which an additional DOF is added to the haptic device by moving its base using a mobile robot, and the position-based workspace drift method is adapted to change the workspace of the HIP in the VE. To prove the efficiency of the proposed workspace increase method, we have conducted a psychophysical stiffness JND experiment to compare its effect on user perception before and after the robotics approach. We have also studied the haptic device's stable and passive operating region for the added robot dynamics.

2 Methodology

2.1 *Design Tools*

For this work, two devices were designed. Initially, a tabletop robot-based device is designed and described as Prototype I, and later a larger multi-axis device is developed and described as Prototype II. While Fig. 1 presents Prototype I, Fig. 2 depicts Prototype II.

2.1.1 Prototype I

The schematic diagram in Fig. 3 explains the main device setup. The robot powered by a 12 V, 2 A DC supply has four motors that run at 300 RPM. The motors were connected to an omni wheel of 100mm diameter. A pair of L293d dual H bridges were used for the bi-directional control of the motors. The motor driver is run by the Atmega328P controller, which gets its input from Unity via a serial port connection. Phantom Omni [2] is used for the haptic feedback, which is placed on the mobile robot of dimension 30 cm × 30 cm. The Phantom device can provide stiffness up to 2.31 N/mm in the 'y' direction. Each motor has an encoder from which the rotation angle is read in the controller. A simple PID control is implemented to precisely update the robot's position. For Virtual Reality (VR) rendering, the HTC Vive system is used due to its large tracking area, and the Vive controllers are used for position detection of the robot. The Vive controllers are position-tracking devices that are connected to a Vive head-mounted display (HMD) wirelessly. Haptic Interface Point

Fig. 1 Prototype I: placement of haptic device over mobile robot which moves only in y-axis

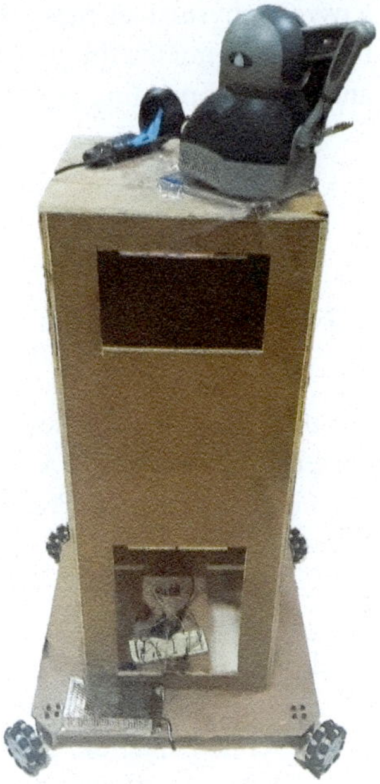

Fig. 2 Prototype II: placement of haptic device on a modified mobile robot which moves in both y and x-axis

(HIP) is the end effector of the phantom haptic device in the Virtual Environment (VE). VE was designed using the Unity 3D game engine, which contains two walls and the virtual haptic tool of HIP. The prototype I was designed only for motion in the 'y' direction.

2.1.2 Prototype II

For the final product, a new robot was designed for easier use. The new robot's base is 60 cm × 60cm, and the dimensions of the high tower are 40 cm × 40 cm with a height of 100 cm. The total weight of the mobile robot is 10 kg, and so a 12 V, 10 A DC power supply powers motors. Four high-gear motors are at the four corners of the new base via an L-clamp to address the robot's overall weight. Each motor can produce a torque of 15 kg-cm at 200 RPM with a stall torque of 40 kg cm, requiring a 7.5 A current to run at full load provided via the motor driver board (rated 24 V and 20 A). These drivers are controlled by an Atmega328P controller, which gets its commands from the Unity VR scene via serial port communication. Each motor has an encoder from which the rotation angle is read in the controller. A simple PID control is implemented to precisely update the robot's position. For VR, the tools used are the same as the prototype I. The haptic device used is a Geomagic Touch mounted at the top of the tower of the robot delivering stiffness up to 2.31 N/mm in the 'y' direction and 1.1 N/mm in the 'x' direction. This new platform is designed for linear motion in the 'x' and 'y' directions, supporting larger-scale VR

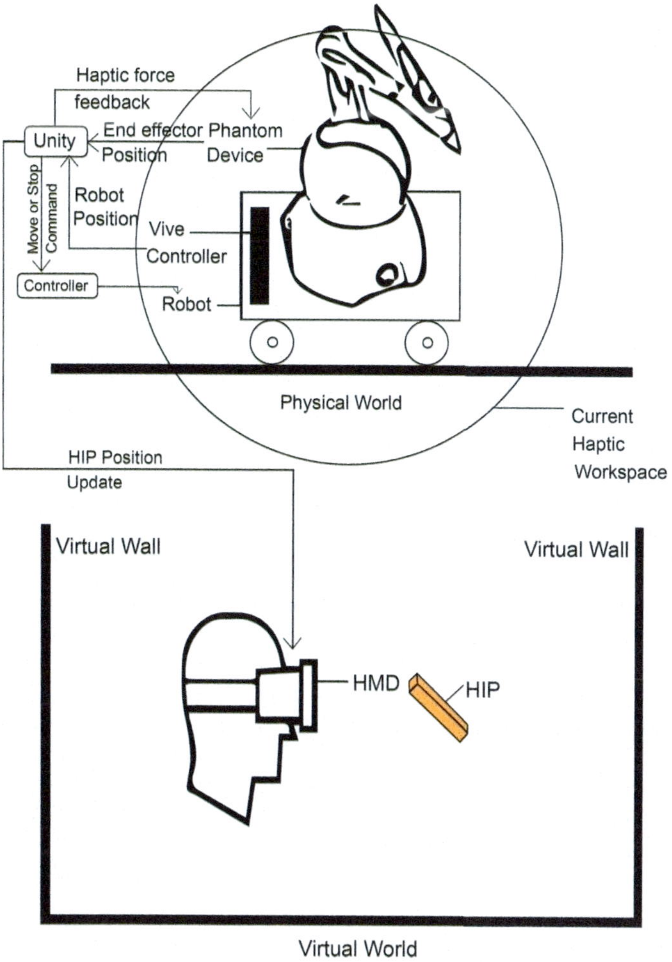

Fig. 3 Schematic diagram of the proposed system that contains the haptic device, mobile robot, user, and HMD

interactions, specifically when the user stands and explores the VE. Even though the larger mobile base adds some inertia, this effect on user interaction is less since the robot is equipped with precise position and velocity control with an update rate of 1000 Hz. Moreover, the recommended mass and damping parameters should be considered in the design of the robot as mentioned in Sect. 2.6.

2.2 Method

The main task involved in this robotics approach is the integration of the haptic device along with the robot's motion in the VE. Since all the Phantom haptic devices are grounded, it is impossible for the user to interact with many objects in the large workspace of VE. Our approach addresses this problem by placing the Phantom Omni on a mobile robot.

Figure 3 shows the overview of the proposed system that contains a haptic device placed on a mobile robot and a user with a Head-Mounted Display (HMD) holding an end effector of the haptic device. In the virtual world, the user can see the simulated virtual walls placed vertically (X = 0) and use the HIP to interact with the walls. Unity gets the position information of the mobile robot from Vive controllers placed over it and the end effector position from the haptic device. The Vive controller is a wireless position-tracking device connected to the Vive HMD system. As the end effector reaches the workspace limit of the Phantom haptic device, the robot moves along the y-axis (right-left), thereby shifting the physical workspace of the haptic device. Figure 4 shows the control architecture for both prototypes I and II. The microcontroller controls each motor through PWM control and reads the angle of rotation from the encoders. The controller transfers the encoder and motor data with the computer via USB communication. At the same time, the HTC Vive controller is connected to HMD via Bluetooth and sends position data to the computer. The haptic device transfers the end-effector position and force data via Ethernet communication to the computer. The robot and haptic device position is updated at the rate of 1000 Hz.

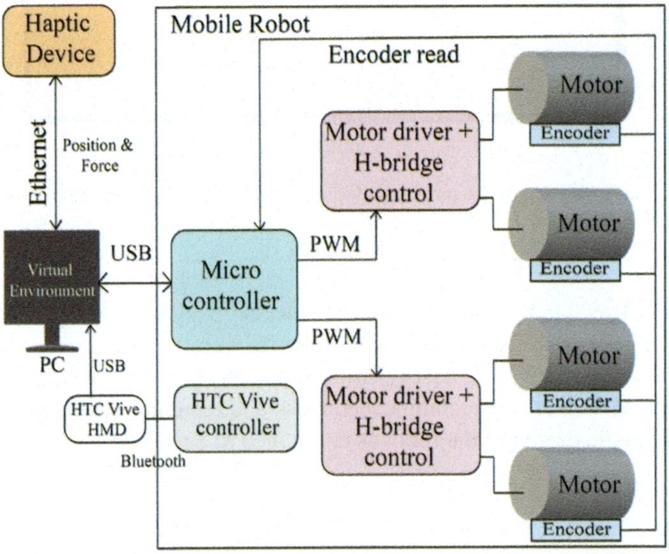

Fig. 4 Control architecture for prototype I and II

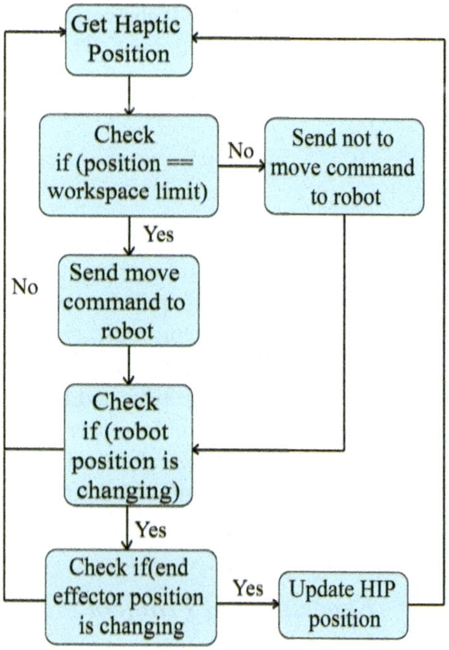

Fig. 5 Flow diagram of position-based workspace drift algorithm; the end-effector position of the haptic device and robot position is continuously checked and the HIP position is updated accordingly

The Workspace Drift Algorithm (WDA) given in Fig. 5 explains the position control algorithm. Through this algorithm, the virtual workspace of the HIP is extended in the VE. Initially, when the user is within the reachable physical workspace of the haptic device, i.e., device workspace, the robot does not move. When the user moves the end effector to the device workspace limit, the robot moves along the axis, and the current position of the HIP is updated based on the positions of the end effector and the robot. Based on the end effector displacement, Unity communicates serially with the controller (ATMEGA328p) for the direction of robot movement. The Vive controller is placed on the robot to track its live position, ultimately providing the distance the robot moves. Based on this, the HIP position is updated as given in (1).

$$VH_P = PWS + EE_P \tag{1}$$

where PWS indicates physical workspace shift, EE_P indicates current end effector position and VH_P denotes Virtual HIP position.

The haptic position in Fig. 5 refers to the end effector position of the haptic device. The end effector position is continuously checked to determine whether it is equal to the device workspace limit. If the end effector is at the workspace limit, then the command to move the robot is given to the motor driver; otherwise, the command to keep the robot in the same position is sent. Then the position of the robot is checked continuously for any change. If the robot is moving and the position changes, then the end effector position is checked. If the end effector position changes, the new

HIP position is updated. If the robot or end effector position is not changing, the haptic position is checked again. These steps are repeated continuously in a haptic rendering loop. The HIP position is updated in real-time for any change in the robot and end-effector position caused by user movement.

2.3 Subjects

Ten healthy subjects participated in Experiment I, and five participated in Experiment II. They had no known neurological and visual perception disorders and no sensory-motor disorders. All subjects were between the ages of 22–29 years with an average weight and height of $75 \pm 10kg$ and $175 \pm 8cm$. All subjects gave informed consent. In Experiment I, seven subjects were right-handed, and three were left-handed. In Experiment II, four subjects were right-handed, and one was left-handed. All subjects performed the Experiment with their dominant hand.

2.4 Experimental Protocol

Two classical psychophysical experiments were conducted in this study. The first experiment was conducted with Prototype I and the next with Prototype II. The experiment was to compare the Just Noticeable Difference (JND) of the stiffness perception in VE during stationary robot cases and moving robot cases. The experiment contains a movable mobile robot that carries the haptic device to interact with the objects in VE, which are far apart. The haptic device was positioned in the center of the robot for better balance. A virtual scene contains two virtual walls on either side of the HIP at equidistant. The user was asked to hold the end effector of the haptic device to move the HIP and explore the stiffness of the virtual wall. The stiffness discrimination task [28] was considered to validate the haptic perception of the user by using a Constant stimuli method [29] from classical psychophysics. In this method, the same set of stimuli with constant intervals are presented in a random order. The task of the subject was to perceive the stiffness of the two virtual walls through haptic feedback and compare them. The paradigm used for this Experiment is Two-Alternative Forced Choice (2AFC) [28]. The subjects were asked to respond "left" if their left virtual wall was stiffer and "right" when the right virtual wall was stiffer. The subjects were allowed to choose "left" or "right" explicitly. Five stiffness levels with a constant interval of 0.345 N/mm were chosen (0.575 N/mm, 0.92 N/mm, 1.265 N/mm, 1.61 N/mm, 1.955 N/mm) as stimuli for the matching task with the reference stimuli of 1.265 N/mm. The stiffness interval was chosen from the pilot study in which the minimum distinguishable stiffness level was identified and the interval was taken less than that stiffness value. This avoids the false estimation of less JND value.

Table 1 Dimensions of the psychophysical experiment

Method	Constant stimuli
Task	Discrimination
Measure	JND
Analysis	Psychometric curve
Stimulus	Stiffness
Paradigm	One interval two alternative force choice

The experiment consisted of two cases: Stationary Robot and Moving Robot. In the stationary robot case, the virtual walls were located within the device workspace and the robot was not moving and in the moving robot case, the virtual walls were located out of the device workspace where the robot had to move to reach the wall. Ten trials were considered for each stiffness level and fifty trials for each case of the experiment. A break of 3 minutes was given between every 25 trials in order to avoid the effect of memory and muscle fatigue. A trial experiment was given to the subjects before the actual experiment to make sure that the haptic cue was used for stiffness differentiation. The trial experiment contains a set of stiffness levels different from the test experiment. The subjects were randomly grouped into two sets for Experiments I and II. In Experiment I, one group (subjects—2, 5, 6, 8, and 9) was asked to perform a stationary robot case followed by a moving robot case and the order was reversed for the other group (subjects—1, 3, 4, 7 and 10). In Experiment II, subjects 2, 3, and 4 (group I) performed stationary robot cases followed by moving robot cases and vice versa for subjects 1 and 5 (group II). This has been followed to cancel out the effect of bias due to the sequential order of the Experiment. The subjects were asked to wear the HMD during the Experiment. In the first experiment, Prototype I was placed on a table at a comfortable distance from the subject and was movable on the table along the y-axis. In the second experiment, Prototype II was placed on the ground and was movable throughout the room on the x and y axes. Table 1 shows the dimensions of the psychophysical experiment that was conducted.

2.5 *Perceptual Analysis*

The stiffness JND was calculated as per the classical psychophysics method as given in [29]. Generally, the psychometric function defines a sigmoid function between the sensory response and the stimuli. The sigmoid function is defined in (2), which has a characteristic "S" shaped curve.

$$S(x) = \frac{1}{1 + e^{-x}} \tag{2}$$

From the experimental data, the proportion of correct response, also called hit rate, is calculated and plotted against the given stimuli from which the JND is derived [30]. JND is a psychophysical measure defined as the minimum amount of change in the stimuli required to perceive the difference. From the psychometric plot, the JND is calculated as given in (3)

$$(\%)JND = \frac{DL_U - DL_L}{2 * St} * 100 \tag{3}$$

For both the cases of the experiment: stationary robot (no change in the workspace) and moving robot (increased workspace). DL_U is the upper threshold, DL_L is the lower threshold and St is the reference stiffness value.

2.6 Stability Analysis

The dynamics of a haptic device are of utmost importance not only because of its structure but also its usability. With the modified dynamics, the physical parameters for the system model shall change. This change in physical parameters will have an impact on the region of stable operation as well as the usability of the device. This involves analysis of stability boundary as well as passivity-based criterion for finding a range of virtual stiffness and damping coefficient. For quantifying the point of interaction displacement in a closed-loop structure, the output function [29] is given in (4).

$$x_h^* = \frac{\mathscr{Z}\{Gf_h\}}{1 + \mathscr{Z}\{ZOH\ G\}C(z)} \tag{4}$$

Where the variables x_h^*, $\mathscr{Z}$, G, f_h, ZOH, and $C(z)$ stand for displacement in the tip of the haptic device, the Z-transform operator, the system transfer function in the Laplace domain, the force exerted on the user's hand, the zero-order hold, and discrete-time controller (virtual environment).

The characteristics equation of a system is given by,

$$1 + \mathscr{Z}\{ZOH\ G\}C(z) = 0 \tag{5}$$

Using the RH criterion for (5) is not straightforward in digital or z-domain. A bilinear transformation from the z-plane to the w-plane helps to analyze the system's stability using the Routh-Hauritz criterion. When expression (6) is satisfied, as stated in [25], the device exhibits stable behavior.

$$\begin{aligned} & b^2KT^2(B+b)e^{-bT/m} \\ & -BbT(mK-b^2)(1-e^{-bT/m})(1+e^{-bT/m}) \\ & +mB(mK-bB-b^2)(1-e^{-bT/m})^2 \\ & -b^2T(mK-b^2)(1-e^{-bT/m}) \\ & +B^2b^2T(1-e^{-bT/m})(e^{-bT/m}) > 0 \end{aligned} \tag{6}$$

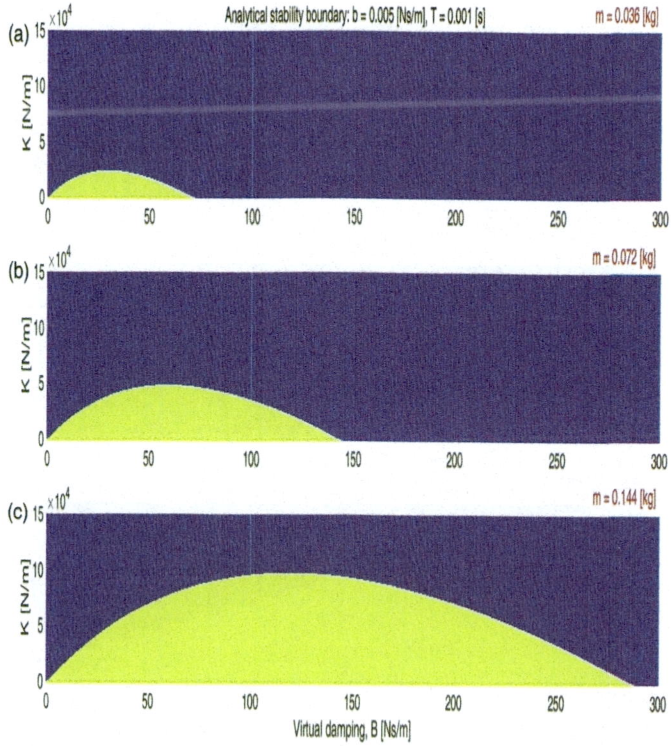

Fig. 6 Analytical stability boundary for a typical haptic device with varied mass. The yellow region denotes the stable operating boundary of the device. **a** For a mass of 0.036 kg, **b** for a mass of 0.072kg, and **c** for a mass of 0.144 kg

The parameters b, m, T, K, and B in the preceding expression denote device damping, mass, sampling time, virtual spring, and virtual damping, respectively. The aim now is to determine the range of virtual environment parameters (K, B) that will allow us to operate the haptic device stably. The parameters for the Phantom haptic device used in this stability analysis were obtained from [26] where the authors studied the stability issues in the context of sampling, delay, and Coulumb effect. Nevertheless, it is essential to examine the stability boundary for the modified haptic device under a fixed sampling rate, and delay, and by neglecting the Coulomb effect. The boundary for stable operation changes with the change in device mass and damping while it remains unaffected by the increase in inertia due to the mobile base. Figure 6 depicts the stable and unstable zones in yellow and magenta, respectively. It is apparent that as mass increases, so does the stable zone.

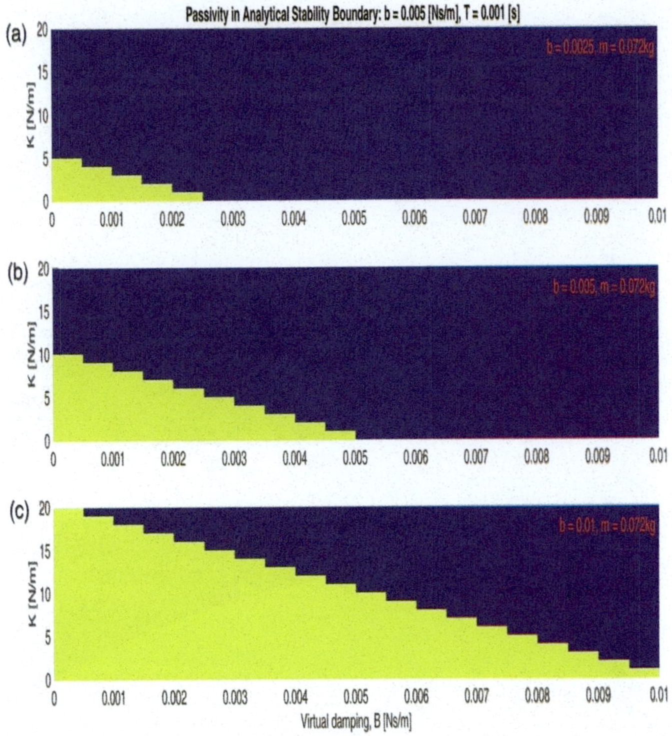

Fig. 7 Analytical passivity boundary for varied physical damping of a typical haptic device. The yellow color lines represent the passive region of the device. **a** For damping of 0.0025 Ns/m, **b** for damping of 0.005 Ns/m, and **c** for damping of 0.01 Ns/m. A passive region is a small region where the device becomes unstable

Moreover, the passivity criterion for sampled data systems [27] applies to this analysis. The passive region and its relation to the physical damping of the device is an important constraint. The expression given in (7) provides the necessary condition for the passivity of a haptic device for stable operation. Figure 7 shows the passive region in yellow color. It can be seen clearly that with the increase in physical damping, the passive region increases and vice versa. The results shown above are very intuitive for understanding the effect of physical parameters on the stable operation of the haptic device and flexibility in K, and B. The noteworthy mention is the intersection of passive and Routh-Hurwitz region of stability, Although a haptic device is stable under the Routh-Hurwitz criterion, it behaves unstably inside the passive region as shown in Fig. 7 which eventually reduces the stable region leading to difficulty in rendering soft-body.

$$b > \frac{KT}{2} + |B| \tag{7}$$

The change in device damping and stiffness also raises the question of what the maximum and minimum values of device parameters in our system model are that assure the device's stability. The passive zone expands as the physical damping value increases, impeding soft body rendering. Similarly, as the physical mass of the haptic device increases, we have unreal interaction with the virtual environment where we have a larger stable region. Although changing the physical characteristics of the haptic device improves system dynamics, it impacts the haptic device's performance, stability, and usability.

3 Results and Discussion

3.1 Experiment I

Figure 8 shows the psychometric function plot for the stationary robot case where there is no robot movement and for the case of the increased workspace involving robot movement. Both the sigmoidal curves are for one subject from which JND was calculated.

The slope of the psychometric curve represents the difference threshold. The mean and variance of the (%) JND have been calculated across 10 subjects for both the conditions of experiment I and given in Table 2. From Table 2, it can be observed that there is only a slight difference in the mean (%) JND between the stationary and robot moving conditions. However, the variance for the robot moving case is more than the stationary case, and it is discussed below.

Figure 9 shows an average psychometric curve for the stationary and robot moving case. From the plot, we can concur that the slope for the two cases is almost equal which means that the robotic approach used in this work has increased the workspace

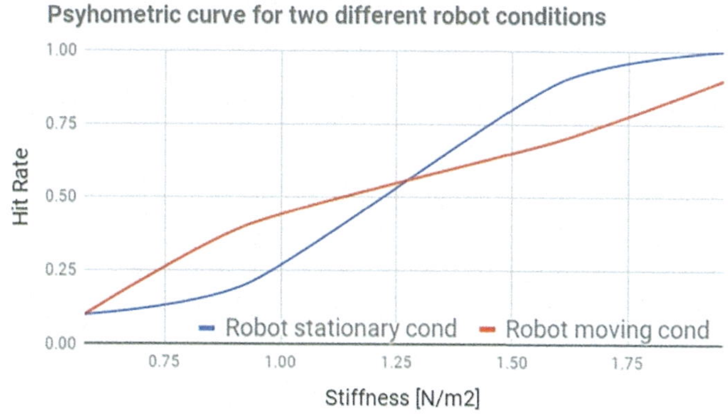

Fig. 8 Psychometric curve for stiffness exploration in VE when the robot is stationary and moving for a typical subject

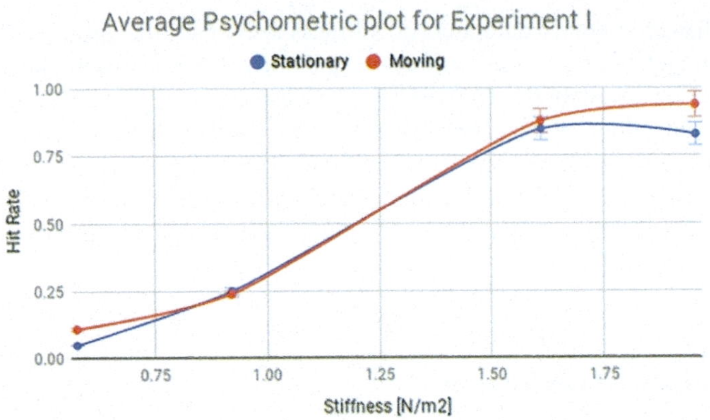

Fig. 9 Average psychometric plot for Experiment I

without affecting the difference threshold of haptic perception. Table 3 shows the average hit rate for the stationary robot case and moving robot case at the four different stiffness levels of experiment I. The hit rate is almost the same for both cases of the experiment across all stiffness levels.

The (%) JND of each subject for both the conditions of experiment I is given in Fig. 10. It can be observed that subjects 1, 3, 4, and 10 (group 1) have slightly lower (%) JND in the robot moving condition than in the stationary condition. Subjects 2, 5, 6, and 8 (group 2) have slightly lower (%) JND for the stationary condition than the

Table 2 Total % JND for experiment I and II

Experimental condition	Experiment I (% JND)		Experiment II (% JND)	
	Mean	Variance	Mean	Variance
Without workspace increase	21.08	22.18	20.55	26.56
With workspace increase	24.59	85.06	26.56	133.92

Table 3 Hit rate for experiment I and II

Stiffness (N/mm)	Experiment I		Experiment II	
	Stationary case	Moving case	Stationary case	Moving case
0.575	0.04	0.09	0.04	0.12
0.92	0.25	0.24	0.20	0.16
1.61	0.85	0.87	0.84	0.72
1.955	0.86	0.93	0.84	0.92

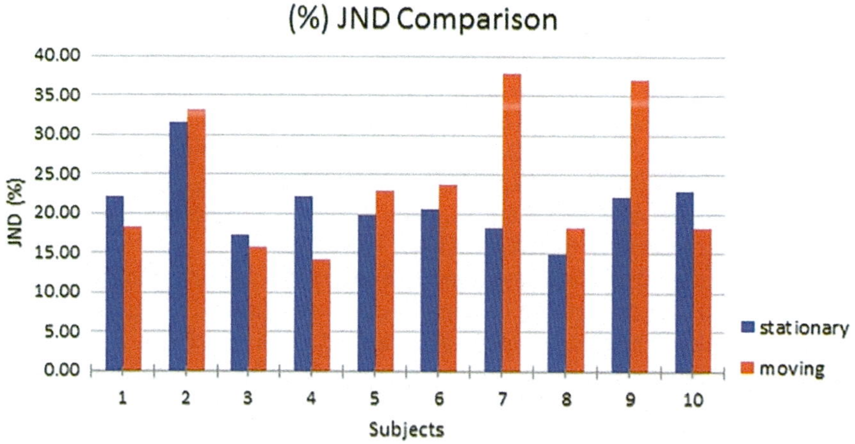

Fig. 10 (%) JND values of stiffness perception of subjects when the robot is stationary and moving for Experiment I

robot moving case. The reason could be the order of the experimental cases performed by the subjects which made them adapted to the experimental task. Even though the trial experiment was given, the subjects were getting more trained in the first case of the experiment which affected their performance in the preceding experimental case. However, since we have two groups of subjects performing different orders, the overall bias effects due to experimental order on the mean (%) JND are reduced. Subjects 7 and 9 are observed to have 15% higher (%) JND for the moving robot case which caused the drastic change in the total variance. The increased distance between the virtual walls also increased the time duration between the stiffness perceptions which might have faded the haptic memory. Although this can be one of the reasons, since other subjects are not showing any memory effect in the experiment, further studies on haptic memory are required to explain this condition.

A one-way ANOVA test has been performed on the experimental data between the robot moving and the stationary conditions. From this test, the observed p-value is 0.32. This indicates that there is no statistically significant difference in the (%) JND between the two conditions of the experiment. This could prove that the proposed method of increasing the haptic device workspace is not significantly affecting the user stiffness perception in the VE.

3.2 Experiment II

A total of five subjects participated in this experiment. Correct responses, wrong responses, and hit rates were noted for each of the stiffness values for all the subjects both in stationary and moving robot cases. Then the psychometric plot was plotted for the two cases for each of the test subjects and their individual %JND was calculated by using Eq. (3). In Experiment II, the experiment protocols are followed and preceded by Prototype II mentioned in the methodology. The mean and variance of the (%) JND have been calculated across five subjects for both the conditions of experiment II and mentioned in Table 2. From Table 2, it can be observed that there is only a slight difference in the mean (%) JND between the stationary and moving robot cases. However, the variance for the increased workspace condition is more than the normal condition, and it is discussed below.

Figure 11 shows an average psychometric curve for the stationary and robot moving case. From the plot, we can concur that the slope for the two cases is almost equal which means that the robotic approach used in this work has increased the workspace without affecting the difference threshold of haptic perception. Table 3 shows the average hit rate for stationary and moving robot cases at the four different stiffness levels of experiment II. The hit rate is almost the same for both cases of the experiment across all stiffness levels.

The (%) JND of each subject for both the conditions of experiment II is given in Fig. 12. It can be observed that subject 2 (group 1) has a slightly lower (%) JND in the moving case than in the stationary case. Subjects 1 and 5 (group 2) have slightly lower (%) JND for the stationary case than the moving case. The reason could be the order of the experimental phases performed by the subjects which made them adapted to the experimental task. Even though the training set was given, the subjects were getting more trained in the first phase of the experiment before performing the second phase. Subject 4 is observed to have a 30% higher (%) JND for the moving robot case which caused the drastic change in the total variance. This could be because of the increase in distance between the virtual walls, which might have faded the haptic memory. Although this can be one of the reasons, since other subjects are not showing any memory effect in the experiment, further studies on haptic memory are required to explain this condition. Also, the high variance could be due to the fewer number of subjects used for experiment II. It is also important to note that Subject 3 is observed to have exactly identical (%) JND in both stationary and moving robot cases which shows that there is no difference between the two cases. The order of the experimental conditions did not play much role, and the learning effect across experimental trials is significantly less for subject 3. The test set was enough to remove the learning effect for subject 3. Also, the prior experience of using gamepads of the subject could be a reason why the subject adapted more quickly to the experiment during the test set than other subjects. Further study with the subjects of having different immersive VE and gaming experiences will explain it.

A one-way ANOVA test has been performed on the experimental data between the moving and the stationary case. From this test, the observed p-value is 0.34. This indicates that there is no statistically significant difference in the (%) JND between the two conditions of the experiment. This proves that the proposed method of workspace expansion does not affect the user stiffness perception in the VE while applying two axes of the haptic device.

In summary, the mean difference (%) JND between the robot moving case and the stationary case is statistically insignificant for both experiments I (y-axis) and

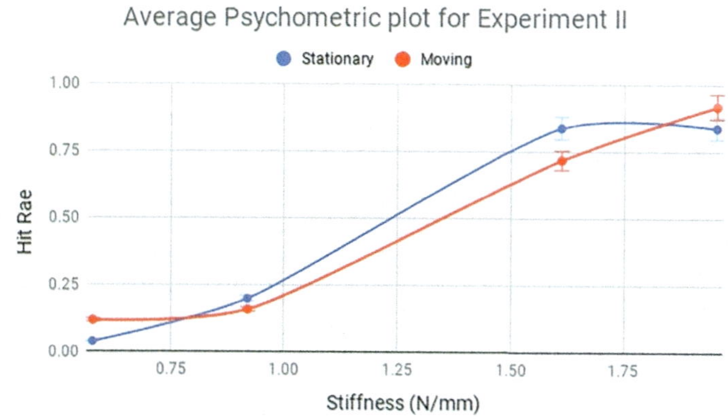

Fig. 11 Average psychometric plot for experiment II

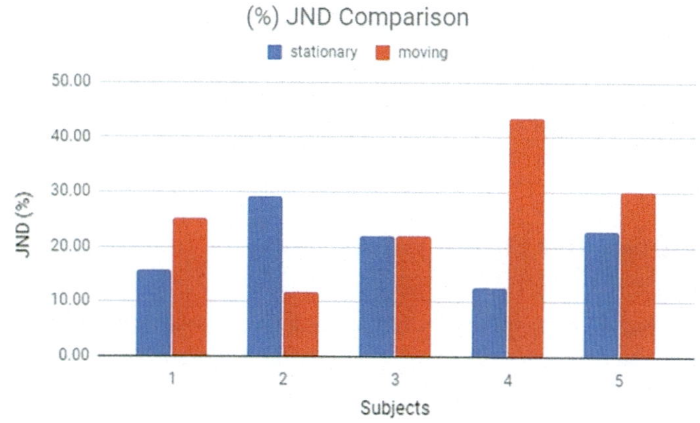

Fig. 12 (%) JND comparison between stationary and moving cases for experiment II

II (x and y-axis movement). The (%) JND of stiffness perception in VE does not vary much during the workspace expansion. This shows that the proposed method of increasing the workspace could not affect the user's stiffness perception, which is the main objective of the current work. When comparing (%) JND for individual subjects between two experimental conditions, (%) JND seems to vary depending on the order of the experimental condition. The (%) JND is slightly higher for the experimental condition performed first. Mild learning is happening over the trials, which is more visible in the experimental condition performed second. The same is observed in both experiments. However, the effect of experimental order is reduced by considering two groups of participants with alternate orders. Another observation is that only for very few subjects, the (%) JND is comparatively higher in the robot moving case than in the stationary robot case in both experiments. One possible reason for this, other than the order effect, could be a haptic memory effect, where the increased distance between virtual walls affected the haptic memory in stiffness comparison during the robot moving case. However, further study on haptic memory and (%) JND will give better clarity on this.

The inclusion criterion for selecting the subjects is no history of any neuromotor disorders. Both genders are included in the experiment. Also, right and left-handed subjects were included in the experiments. The results found are for the age group (22–29 years) mentioned in the paper. For elderly subjects, the range of (%) JND for both the experimental conditions and haptic memory effect may vary, but the mean difference (%) JND may still be the same. However, additional experiment is needed to confirm it. Also, since there is no dependency between the haptic device and the robot, the proposed technique can be used for any grounded haptic device with proper fixing to the robot. The current method is implemented with up to two axes but can be extended to three. In the present study, the distance between the two virtual walls is slightly larger than the workspace of the haptic device during the robot moving case. The distance between two virtual walls can be varied, and that effect on the mean (%) JND must be studied to analyze the proposed method's efficiency further. Also, in the current study, the virtual walls are stationary objects, and only the robot was moving. However, the proposed method has to be tested for dynamic virtual objects.

4 Conclusion and Future Works

The main objective is to increase the haptic device workspace by using a mobile robot and position-based workspace drifting algorithm and to check whether the user perception is affected. This method avoids the issues reported in the literature for workspace expansion such as the usage of fixed redundancy axes and control algorithms that only vary the virtual tool workspace through scaling or drifting. Since the given method shifts the workspace simultaneously in physical and virtual environments, it can render haptic feedback without any interruption while moving the haptic device. The proposed method has been analyzed through a psychophysical study for stiffness exploration in the virtual environment. From the study, the stiffness (%) JND for expanded workspace through the implemented method is found to be almost the same as in the initial workspace of the haptic device. This could prove that the method is effective in increasing the workspace area of the haptic device without affecting the users' haptic perception. The effect of the robot on the haptic device performance is analyzed through the stability of the system. From the analysis, the possible system parameters can be chosen.

Although current work is limited to two axes, this methodology can be extended to other axes and any grounded haptic devices. Visual Simultaneous Localization and Mapping (VSLAM) can be integrated to avoid obstacles and plan the trajectory to have an optimized and robust control of the robot. Also, further studies on haptic memory will explain the observed variations in the conducted experiment. Studies with different psychophysical metrics such as force JND and response time will help to analyze the effectiveness of this approach further.

References

1. Barrow A, Harwin W (2016) Design and analysis of a haptic device design for large and fast movements. Machines 4:8. https://doi.org/10.3390/machines4010008
2. Silva AJ, Ramirez OAD, Vega VP, Oliver JPO (2009) Phantom omni haptic device: kinematic and manipulability. In: Electronics, robotics and automotive mechanics conference, CERMA'09, IEEE, pp 193–198
3. Puerto MJ, Sánchez E, Gil JJ (2009) Control strategies applied to kinesthetic haptic devices. https://doi.org/10.1109/RIISS.2009.4937918
4. Giri GS, Maddahi Y, Zareinia K (2021) An application-based review of haptics technology. Robotics 10:29. https://doi.org/10.3390/robotics10010029
5. Ramstein C, Hayward V (1994) The pantograph: a large workspace haptic device for a multimodal human-computer interaction. In: Conference on human factors in computing systems ACM/SIGCHI companion-4/94 (CHI'94), pp 57–58
6. Gosselin F, Andriot C, Savall J, Martín J (2008) Large workspace haptic devices for human-scale interaction: a survey. In: Ferre M (ed) Haptics: perception, devices and scenarios, EuroHaptics, vol 5024. Lecture Notes in Computer Science. Springer, Berlin, Heidelberg
7. Dominjon L, Perret J, Lécuyer A (2007) Novel devices and interaction techniques for human-scale haptics. Visual Comput 23:257–266
8. Prattichizzo D, Chinello F, Pacchierotti C, Malvezzi M (2013) Towards wearability in fingertip haptics: a 3-DOF wearable device for cutaneous force feedback. IEEE Trans Haptics 6(4):506–516
9. Laycock SD, Day AM (2003) Recent developments and applications of haptic devices. In: Computer graphics forum, vol 22(2). Blackwell Publishing, Inc, Oxford, UK
10. Lee G, Hur S, Oh Y (2016) A novel haptic device with high-force display capability and wide workspace. In: IEEE international conference on robotics and automation (ICRA). Stockholm, pp 2704–2709
11. Basdogan C, Srinivasan MA (2001) Haptic rendering in virtual environments
12. Conti F, Khatib O (2005) Spanning large workspaces using small haptic devices. In: First joint eurohaptics conference and symposium on haptic interfaces for virtual environment and teleoperator systems, World Haptics Conference. Pisa, Italy, pp 183–188
13. Garlington I (2012) Expanding the usable workspace of a haptic device by placing it on a moving base
14. Cruz-Neira C, Leigh J, Papka M, Barnes C, Cohen SM, Das S, Vasilakis C et al (1993) Scientists in wonderland: a report on visualization applications in the CAVE virtual reality environment. In: Proceedings of 1993 IEEE research properties in virtual reality symposium, IEEE, pp 59–66
15. Chotiprayanakul P, Liu DK (2009) Workspace mapping and force control for small haptic device based robot teleoperation. In: International conference on information and automation. Zhuhai, Macau, pp 1613–1618
16. Dominjon L, Lecuyer A, Burkhardt JM, Andrade-Barroso G, Richir S (2005) The "Bubble" technique: interacting with large virtual environments using haptic devices with limited workspace. In: First joint eurohaptics conference and symposium on haptic interfaces for virtual environment and teleoperator systems, world haptics conference. Pisa, Italy, pp 639–640
17. Mamdouh M, Ramadan AA, Abo-Ismail AA (2012) Evaluation of a proposed workspace spanning technique for small haptic device based manipulator teleoperation, pp 161–170
18. Salcudean SE, Wong NM, Hollis RL (1995) Design and control of a force-reflecting teleoperation system with magnetically levitated master and wrist. IEEE Trans Robot Autom 11(6):844–858

19. Pascale M, Formaglio A, Prattichizzo D (2006) A mobile platform for haptic grasping in large environments. Virtual Real 10(1):11
20. Formaglio A, Giannitrapani A, Franzini M, Prattichizzo D, Barbagli F (2005) Performance of mobile haptic interfaces. In: Proceedings of the 44th IEEE conference on decision and control, IEEE, pp 8343–8348
21. Formaglio A, Prattichizzo D (2005) Smooth approximation of mobile platform displacement for mobile haptic interfaces. In: Proceedings of 2nd international conference on enactive interfaces
22. Nitzsche N, Hanebeck UD, Schmidt G (2003) Design issues of mobile haptic interfaces. J Robot Syst 20(9):549–556
23. Lee I, Hwang I, Han KL, Choi OK, Choi S, Lee JS (2009) System improvements in mobile haptic interface. In: World haptics 2009-third joint eurohaptics conference and symposium on haptic interfaces for virtual environment and teleoperator systems. IEEE, pp 109–114
24. Gourishetti R, Manivannan M (2018) Improved force JND in immersive virtual reality needle insertion simulation. Virtual Real. https://doi.org/10.1007/s10055-018-0369-9
25. Gil JJ, Avello A, Rubio A, Florez J (2004) Stability analysis of a 1 DOF haptic interface using the Routh-Hurwitz criterion. IEEE Trans Control Syst Technol 12(4):583–588
26. Diolaiti N, Niemeyer G, Barbagli F, Salisbury JK (2006) Stability of haptic rendering: discretization, quantization, time delay, and coulomb effects. IEEE Trans Rob 22(2):256–268
27. Colgate JE, Schenkel GG (1997) Passivity of a class of sampled-data systems: application to haptic interfaces. J Robot Syst 14:37–47
28. Maher C, Adams R (1995) A psychophysical evaluation of manual stiffness discrimination. Aust J Physiother 41(3):161–167
29. Gescheider GA (1997) Psycophysics: the fundamentals, 3rd edn. Lawrence Erl-baum Associates Inc
30. Gurari N, Kuchenbecker KJ, Okamura AM (2009) Stiffness discrimination with visual and proprioceptive cues. In: World haptics 2009—third joint Eurohaptics conference and symposium on haptic interfaces for virtual environment and teleoperator systems. Salt Lake City, UT, pp 121–126

Multibody Dynamics Software-Based Simulation of a Game for a Robotics Competition

Sandeep Kumar, Subir Kumar Saha, and Satinder Paul Singh

Abstract Due to advancements in the area of robotics and automation, there is a need of skills for engineers. They are obtained through various workshops, seminars, conferences and participating in various competitions. Apart from the practical work skills, there is a requirement of proper software-based skills to get an idea of the performance of final tasks on the field. This paper presents the use of such a multi-body dynamics software for simulating the game of a robotics competition to check the effectiveness and completeness of the intended tasks. In this work, MSC ADAMS, a multi-body dynamics software, has been used for simulations. The tasks that have been performed are the tasks for ABU Robocon 2022. The theme was "Lagori" that is based on Indian game called as "Pithu". Two tasks of the game were considered here to demonstrate the use of a software. One is Lagori breaking and the other is hitting the Ball on Head. The simulation was performed for several speeds and angles at which the ball was thrown. The effectiveness of the tasks was checked with the real situations when the games were actually played.

Keywords Lagori · Simulation · ADAMS · Robocon

1 Introduction

Pithu/Lagori is an Indian traditional game as depicted in Fig. 1 [1]. The game is played between two teams and the game starts by throwing a ball by the Seeker team to break a stone tower or discs pyramid called "Lagori". While the Seekers try

S. Kumar (✉) · S. K. Saha · S. P. Singh
Department of Mechanical Engineering, Indian Institute of Technology Delhi Hauz-Khas, New Delhi 110016, India
e-mail: mez198423@mech.iitd.ac.in

S. K. Saha
e-mail: saha@mech.iitd.ac.in

S. P. Singh
e-mail: singhsp@mech.iitd.ac.in

S. K. Saha and K. Rama Krishna (eds.), *Recent Developments in Multibody Dynamics*, Lecture Notes in Mechanical Engineering,
https://doi.org/10.1007/978-981-97-7525-5_14

Fig. 1 Pithu game (Lagori)

to pile up the stones again, the hitter team throw ball to interrupt them. There are researchers who applied the engineering knowledge in sports. A impact dynamics analysis of a tennis ball with the flat surface by considering the spring damper model in vertical direction and rotary motion separately [2]. Similarly impact analysis on tennis racket/ball was done [3]. In this, tennis racket was modeled as nine-point masses connected through 24 springs. A three-dimensional motion of a cricket ball was done using volumetric contact model and it was tested for leg break and top spin ball [4]. A number of simulations to study the effect of the badminton serve release height were done [5]. These and similar research are the motivation behind the work of the current paper. The game in this case is Pithu or Lagori.

Now the game of Lagori was adopted as the theme of ABU Robocon, which was organized by India and hosted in New Delhi on August 21, 2022. In this competition, the students need to prepare two robots to complete some tasks. Some of the tasks are like: Lagori breaking, Lagori piling, Pickings balls and hitting the ball to displace it [6]. Out of these tasks, Lagori breaking and Hitting the ball on the head of other robots to displace it well qualify for multibody dynamics simulation and analysis. To perform these tasks properly on the field, it is wiser to simulate the problems using a simulation software, say RecurDyn, ADAMS or similar. In this paper, ADAMS [7] student version was used, which was freely available to the first author who performed all the analysis. Simulating the problem before doing the tasks on the field enhances the skill of the students and gives them confidence. This paper shows various simulations varying the different parameters which may help organizers to fix the game plan and students to fabricate the proper robots to perform the tasks effectively and efficiently. In this paper, the simulations were performed to check the feasibility of various tasks and rules prescribed in the rulebook. Later, they were verified with real data captured during the actual game. In the next section of this paper, methodology is explained. Then results are presented for different scenarios and finally the paper ends with the conclusions.

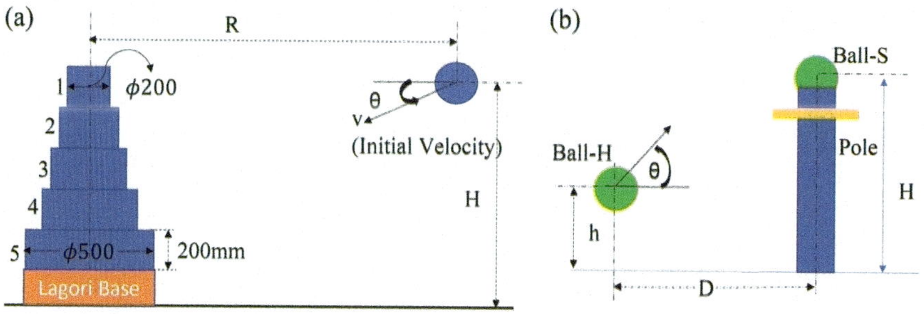

Fig. 2 **a** A ball hitting the Lagori (made of 5 discs) **b** A ball hitting another ball placed on top of a robot

2 Methodology

Pithu/Lagori is a game of breaking the pyramid made of discs placed one over the other as shown in Fig. 2a by a team called "Seeker", and piling up the pyramid without getting hit by the ball of opponent's team called "Hitter". The Hitter team needs to throw a ball and hit the ball placed over the top of a robot of Seeker team who is piling up the pyramid. Figure 2b illustrates the situation where "Pole" simulates the seeker team's robot carrying a ball indicated as "Ball-S", whereas hitter team's ball is represented as "Ball-H".

The above two tasks are simulated in this work ignoring the spinning of the ball and air drag forces. The variables considered in the first set of simulations were speed of the ball, height (H) and placement of the discs, etc. The current arrangement shown is 54321, here 5 means the largest disc with diameter 500mm, 4 means the second largest with diameter 425 mm, 3 means the third largest with diameter 350 mm, 2 means the fourth largest with diameter 275 mm and 1 means the smallest Lagori disc with diameter 200 mm. The balls have a diameters of 140mm. The contact model was implemented in all the possible contacts. The ball is made of leather and discs are made of polyurethane foam. The ground was considered to be made of wood. The contacts in the first task were disc-floor, disc-disc, disc-ball and ball-floor. The contact model was based on the Hertzian model written as $F = k \times x^e$, here k is the contact stiffness, e is the force exponent and x is the penetration depth. Materials properties used are given in Table 1. All the parameters related to contact model for the simulations are taken close to those available online, which are given in Table 2.

Table 1 Material properties of objects used in simulations

Object	Ball	Lagori disc	Floor
Material	Leather	Polyurethane foam	Wood
Density (Kg/m^3)	1200	14	680
Modulus of elasticity (E)	40 MPa	15 MPa	8 GPa
Poisson ratio	0.4	0.33	0.3

Table 2 Contact properties in the Hertzian model

Contact type		Disc-floor contact	Disc-disc contact	Disc-ball contact	Ball-floor contact
Stiffness (N/mm)		100	1000	100000	10000
Force exponent		2.2	2.2	2.2	2.2
Damping		10	10	10	10
Penetration depth (mm)		0.1	0.1	0.1	0.1
Coefficient of friction	Static	0.4	2.95	0.2	0.3
	Dynamic	0.2	1.95	0.08	0.1

Note that, static and dynamic friction coefficients were used for all contacts. Damping was used to create the damping force that dissipated energy. These kinds of simulations are also very important for understanding the physics of the tasks to be carried out during the competition. These simulations checked the angle, speed and height of the ball to be thrown to break the Lagori so that maximum points can be earned.

For the second case, the target was to drop the ball placed on the top of the robot by hitting at an appropriate point. Similar to the first case, various simulations were done by varying the various parameters like speed, angle, height (h) of the ball thrown from the ground and height (H) of the ball placed at top of the Seeker robot. The initial speed and angle were calculated using projectile motion. This paper gives an overview of the task simulations that were needed to check the feasibility of the game rules planned by organizers but same could be performed by any team before their robots are built for the competitions.

3 Results and Discussion

Various simulations were performed to check the effectiveness of Task 1, with parameters shown in Table 3. In all these simulations, $R = 3$ m, simulation time was kept as 3 s with number of steps as 1000. In all the simulations, 4 lagoris were broken, except the case when H $=$ 1 m, v $=$ 20m/s and $\theta = 45°$.

Various Lagori combinations were simulated, like 54321, 45321 and 34521 shown Fig. 3a–c, respectively. In these cases, the distance between throwing position and current position $=$ 2.47m, Speed of ball $=$ 14.82 m/s, Angle of projection (angle between ball path and horizontal Line) $\theta = 13.6$ deg, as shown in Fig. 3a. It has been

Table 3 Parameters for various simulations of Lagori breaking

h	1 m				0.5 m			
v	20 m/s		12 m/s		20 m/s		12 m/s	
θ	0°	45°	0°	45°	0°	45°	0°	45°

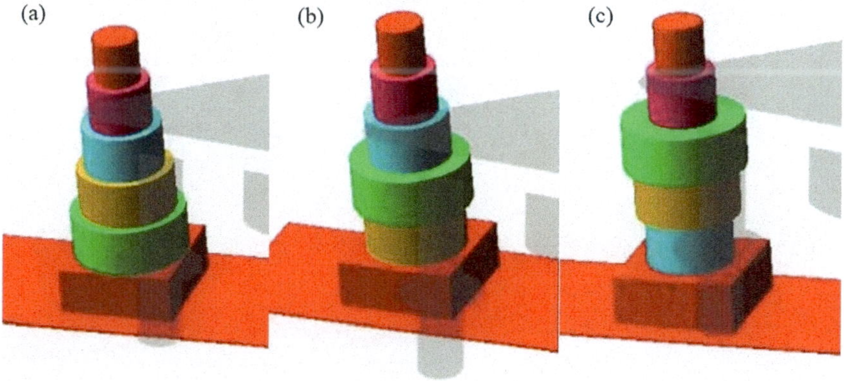

Fig. 3 Combination of lagoris **a** 54321 **b** 45321 **c** 34521

observed that with the combination 45321, only three lagori disc were broken while in others two combinations, all the lagori discs were broken. Finally, in the end, 54321 combination of lagori discs was chosen for ABU Robocon 2022 and what has been observed is that when ball hits the largest lagori disc, i.e. 5, the lagori breaking was successful. The result of hitting the lagori discs with ball can be seen in Figs. 4, 5 and 6. The observations during the actual games played on 16–17 July at Thyagraj stadium, New Delhi, India for the national championship (DD-Robocon) are also shown in each Figure. It is clear from Fig. 4, where ball is hitting at the largest lagori disc, the possibility of breaking the lagori is quite high. The friction between the lagori base and the largerst lagori disc is low due to Sunmica layer pasted on the bottom of the largest lagori disc. In Fig. 5, where ball is hitting the middle lagori, i.e. 3, the possibility of breaking the complete lagori stack is low which is visible in both the simulation and in the actual game. In Fig. 6, where ball hits the base which is fixed, it is obvious that, this will not be able to break the lagori but this can give the idea about the lowest value of speed when the ball will not hit any lagori disc but touches lagori base. Same observation has been observed actual game also.

In the second case where Ball-H hits the Ball-S, first attempt was made using the parameters given in the Table 4. Case was solved by making it a optimization problem, which is formulated in Eqs. (1) and (2). Where θ and u being the angle of Ball-H thrown with horizontal and speed, respectively. Here D and H were calculated as per ABU Robocon Online rulebook.

$$Minimize\ function\ \ |f(\theta)| = h - D\tan(\theta) + \frac{9.81 \times 7.2111^2}{2u^2\cos^2(\theta)} \tag{1}$$

Subjected to the constraints:

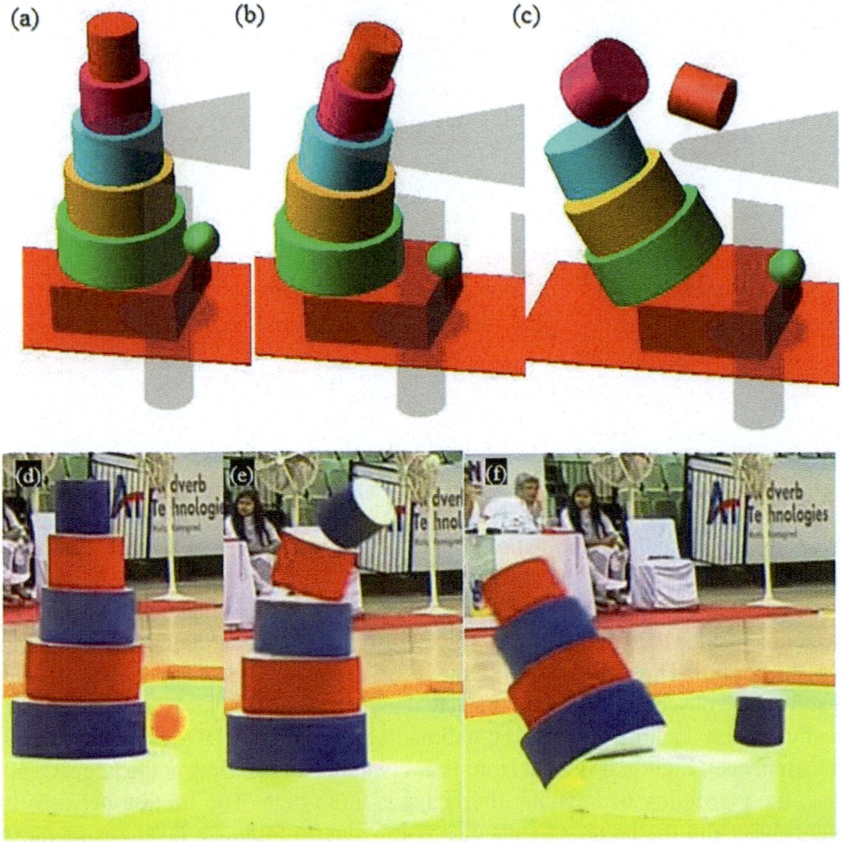

Fig. 4 Ball hitting at the largest lagori at time steps of **a** 0.1500 s, **b** 0.2100 s, **c** 0.4028 s and **d–f** During DD-Robocon 2022

$$
\begin{aligned}
0 \leq h &\leq 1.3, \\
u &\leq 9, \\
0 \leq \theta &\leq \pi/2
\end{aligned}
\tag{2}
$$

It has been observed that when speed (u) was 8.866 m/sec, the target ball "Ball-S" was dropped due to hitting while it was not dropped when speed (u) was 9 m/sec, because the Ball-H hit on the top of Ball-S and it comes back to the initial position. In these cases, internal diameter of the pipe was 110 mm. In the second attempt, internal diameter of pipe was varied and kept as 60 mm, 90 mm and 110 mm as shown in the Fig. 7a–c, respectively. In these attempts, parameters were as follows: Distance between throwing position and the current position (D) = 1.612 m, Speed of ball (u) = 8.866 m/sec, Angle of projection (angle between ball path and horizontal line) θ = 7.125 deg. In all the cases, Ball-H was successfully able to drop the Ball-S.

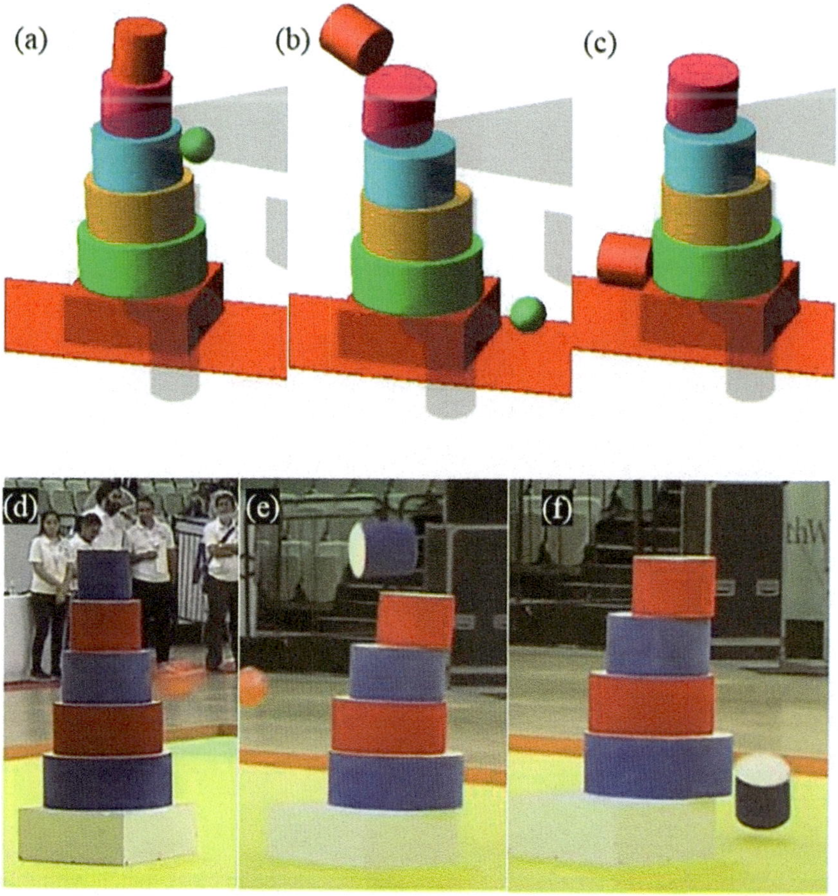

Fig. 5 Ball hitting at the middle lagori at the time steps of **a** 0.1500 s, **b** 0.3075 s and **c** 0.4888 s **d–f** During DD-Robocon 2022

Possibility of dropping the target Ball-S from platform is less if hitting Ball-H hit at the top of target Ball-S. Finally, ball was placed on the pipe of internal diameter 100mm. The results for the same are shown in Fig. 8.

4 Conclusions

This paper provides the various simulations of Lagori breaking, and hitting the "Ball-S" which is on head of Seeker's robot played during ABU Robocon 2022. In case of Lagori breaking, with Lagori combination 54321, 4 lagori disc were broken, except the case of $H = 1$ m, $v = 20$ m/s and $\theta = 45°$. It has been observed that with 45321, only three lagoris are broken while with others two combinations, all the lagori

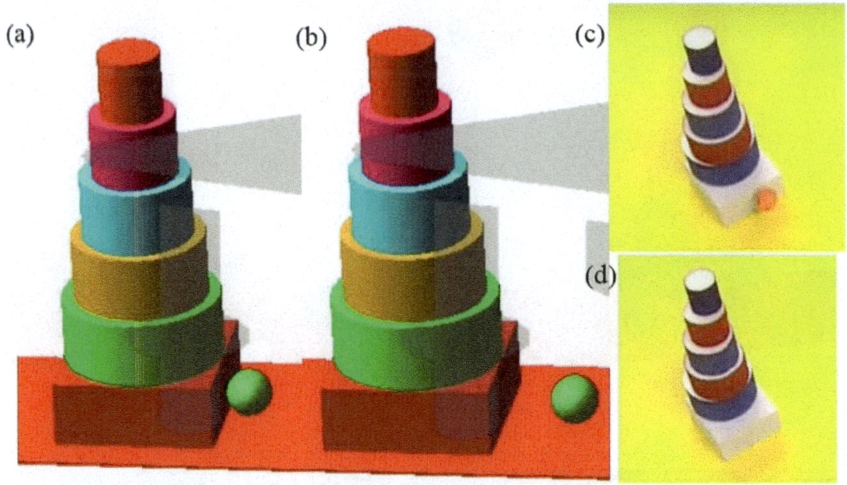

Fig. 6 Ball hitting at the lagori base at the time steps of **a** 0.4440 s **b** 0.6570 s **c–d** During DD-Robocon 2022

Table 4 Parameters for ball hitting ball on head

Height (H) = 1.288 m	
D = 4.1231 m	D = 7.2111 m
h = 1.3 m	h = 1.3 m
u = 7.266 m/s	u = 7.266 m/s

were broken. Some of the observations were found to be in close match with those found during the actual Robocon matches. In case of "Ball-H" hitting "Ball-S", it has been observed that when speed was 8.866 m/s, the target ball was dropped hue to hitting while it's not dropped when speed was 9 m/s because ball hit on the top of it and it comes back to the pipe with internal diameter 110 mm. By varying the hole internal diameter, in all cases, ball is successfully able to drop the ball. In order to simulate the combinations of u, h, and D optimization formulation was formed. Then, simulation with the actual distance and height of BOH is shown with three cases using optimization problem. The contributions of the paper can be listed as follows.

1. Use of advanced simulation software like ADAMS for the game planning of a robotics competition, namely, ABU-Robocon 2022 hosted by India.
2. Optimization of the game parameters.
3. A framework to illustrate how modeling environment can be established in the similar planning of the game rules.

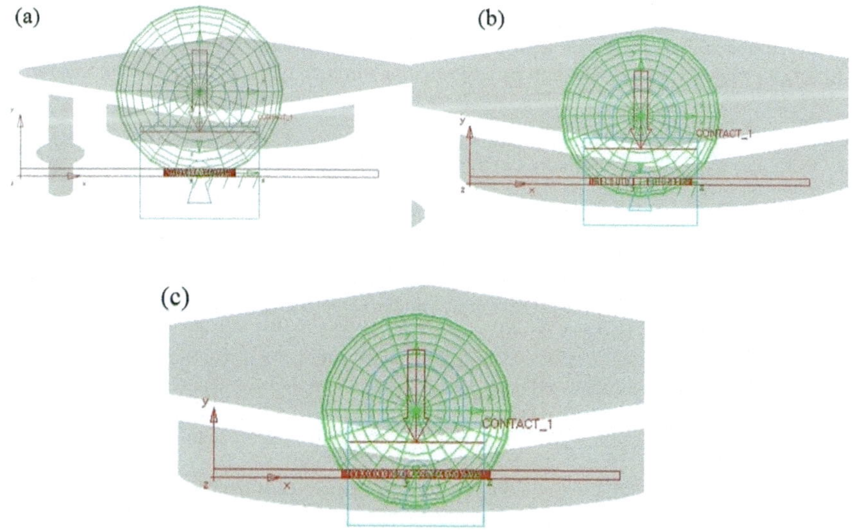

Fig. 7 Pipe internal Diameter **a** 60 mm **b** 90 mm **c** 110 mm

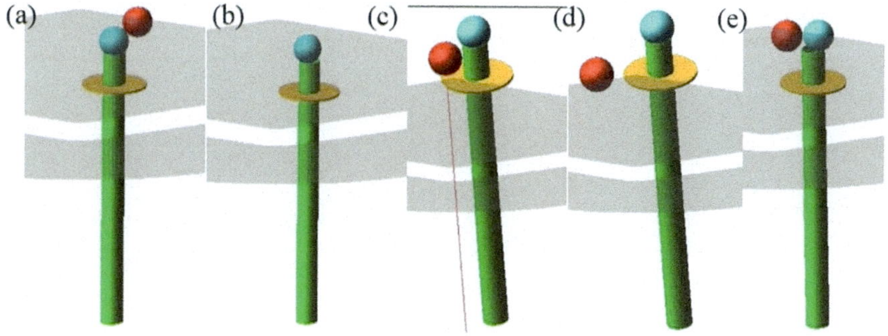

Fig. 8 **a–b** Ball touching at top, (unsuccessful) **c–d** ball hitting pipe, (unsuccessful) **e** Ball hitting ball, (successful in displacing the ball on head (BOH)

4. The established understanding of the Lagori game's dynamics opens up possibilities of utilizing the obtained insights as initial conditions for the development of robotic systems capable of playing Lagori.

As a future scope, these simulations can enable a better understanding of games by incorporating more realistic situations, such as considering factors like ball spinning and air drag force. These simulations can be utilized to develop video games based on physics, not only for Lagori but also for other games. By leveraging these simulations, developers can create more immersive and authentic gaming experiences that take into account the complexities of real-world physics, enhancing the overall gaming experience for players.

Acknowledgements Authors sincerely acknowledge the organizing committee of ABU-Robocon and DD-Robocon 2022 for providing the game data and formation of the game rules. The authors also acknowledge the Prasar Bharti team for providing the good quality videos of the competition.

References

1. Simple Days Lagori Pitthu handmade wooden traditional indian seven stone outdoor game—Amazon.in. (n.d.). https://www.amazon.in/Simple-Days-Handmade-Traditional-Outdoor/dp/B07WGGWFHG?ref-=ast-sto-dp. Accessed 13 Oct 2022
2. Mohsin Jafri SM, Vance J (2005) Modeling of impact dynamics of a tennis ball with a flat surface. In: International design engineering technical conferences and computers and information in engineering conference, pp 399–412
3. Glynn J, King M, Mitchell S (2011) A computer simulation model of tennis racket/ball impacts. Sports Eng 13(2):65–72
4. Banerjee J, McPhee J (2012) Simulation of the three-dimensional motion of a cricket ball using a volumetric contact model. Proced Eng 34:676–681
5. Rasmussen J, Zee M (2021) A simulation of the effects of badminton serve release height. Appl Sci 11(7):2903
6. ABU Robocon (2022) Rulebook. https://www.aburobocon2022.com/index.php/game-rules. Accessed 13 Oct 2022
7. ADMAS student edition (2021) User manual

Modelling and Diagnosis of Defects in Spur Gear Under Constant Speed Operation

Rajeev Kumar, Ranjan Gouda, Samrat Mandal, and Chintamani Mishra

Abstract Gearbox is frequently utilized as a power transmitting device due to its adaptability to operate at a wide range of speed and loads. Any gearbox defect may lead to total failure or even seizure of the machine. Therefore, predictive maintenance is adopted to find the health state of gear train. Modelling of a physical system is economical due to the cost involved in experiments. In the current work, a dynamic model of 9-degrees of freedom (DOFs) spur gear model is taken into consideration to determine the status of gear in both healthy and faulty conditions. In modelling, a high contact ratio spur gear system is considered for analysis. The model comprises of a loader, a motor, and a gear-pinion pair. The equations of motion from the planer block diagram model of the gear pinion system are solved in the MATLAB-Simulink environment. Time and frequency domain analysis is used to compare the health of gear train under different operating conditions.

Keywords Gear mesh frequency · Modelling and simulation · Vibration analysis · Condition monitoring · Single stage spur gear

1 Introduction

Gearbox is one of the most useful elements to transmit power in several rotating machinery, such as automobile, aerospace, wind power, miming machinery, etc. Different types of faults like pitting, spalling, fatigue, broken teeth, tooth crack, etc. are common in case of gearbox after a long cycle of work. These faults may cause catastrophic failure, economic losses and casualties. To avoid such failure, condition-based

Rajeev Kumar, Ranjan Gouda, Samrat Mandal and Chintamani Mishra are the authors contributed equally to this work.

R. Kumar · R. Gouda · S. Mandal · C. Mishra (✉)
Department of Mechanical Engineering, NIT Durgapur, Durgapur 713209, West Bengal, India
e-mail: cmishra@me.nitdgp.ac.in

S. Mandal
e-mail: sm.20me1104@phd.nitdgp.ac.in

S. K. Saha and K. Rama Krishna (eds.), *Recent Developments in Multibody Dynamics*, Lecture Notes in Mechanical Engineering,
https://doi.org/10.1007/978-981-97-7525-5_15

monitoring technique can be used. Out of several condition monitoring techniques, vibration-based condition monitoring technique is adopted in this study. In this technique, the vibration characteristics of the system are studied for identification of fault. To reduce experimental cost and wastage of materials, numerical modelling as well as finite element method (FEM) is used for the dynamic modelling of the gear transmission system. In this study, the pinion having a increasing order of broken teeth is considered as fault in the gear system. In order to analyse the time varying mesh stiffness (TVMS) of both healthy and defective gears, numerous researchers have devised analytical methods as well as finite element method (FEM). Analytical, as well as FEM-based methods for calculating TVMS, has some merits as well as demerits. FEM-based methods give more accurate results as compared to the analytical method [1, 2] but it is complicated and has low computational efficiency whereas the analytical-based method is more ease to use and has high computational efficiency [3–5]. Yang et al. [6] used the potential energy (PE) based method to determine the TVMS of a cracked gear. In modelling, they considered the effects due to hertzian, bending, and axial compressive energies. Tian [7] analysed the impacts of chipped, broken, and cracked teeth on TVMS by taking into account shear energy and compared the response of gear system to that of PE based approach as proposed by Yangs et al. [6]. Based on Tian's [7] model, Wu et al. [8] developed a mathematical model of a single-stage spur gear system for calculating TVMS under different levels of tooth root crack and also studied the effect of tooth crack growth on the vibration response of gear set. In order to examine the impact of gear tooth crack on the TVMS, Chen et al. [2, 9] devised an analytical model by considering crack propagation along the tooth width and depth simultaneously. Pandya et al. [10] studied the crack propagation path of high contact ratio ($C > 2$) for spur gear system, having crack on tooth root using FEM method. Mohammed et al. [11] performed a comprehensive review on modelling of single stage spur gear system, where modelling of gear with six, eight and twelve DOFs are discussed in detail. Kumar et al. [12] developed two models of bevel gear system to study its response with increasing level of fault. Several researchers studied the vibration response of dynamic equipment to know the health status. Fault-induced signal is generally corrugated with noise. Therefore, the signal is pre-processed to remove the noise with help of some of the technique like wavelet analysis [13–15], the empirical model decomposition (EDM) [16], etc. For analysing the postprocessed signal, some of the widely used techniques are spectral kurtosis [1, 2], root mean square (RMS) value analysis [17], and the autoregressive (AR) model [18], etc. In the present work, a PE-based analytical method [2] is used for calculating TVMS of healthy as well as faulty gear train and the TVMS value is used in 9-DOF spur gear system to compute the vibration response of high contact ratio spur gear system under healthy as well as faulty conditions.

2 Calculation of Equivalent Stiffness of Spur Gear System

The equivalent stiffness of spur gear system in both healthy and faulty conditions is calculated by the PE-based method proposed by Chen et al. [2]. In this method, gear tooth stiffness is calculated as a function of shear energy, bending energy, axial compressive energy, Hertzian energy and fillet foundation deflection. In this method, angular variables are used instead of linear variable for individual stiffness calculation.

2.1 For Healthy Condition

The Hertzian stiffness (k_h), bending stiffness (k_b), shear stiffness (k_s) and the axial compressive stiffness (k_a) are calculated as per the Eqs. 1 to 4, respectively [4].

$$k_h = \frac{\pi EL}{4\left(1-\nu^2\right)} \tag{1}$$

$$\frac{1}{k_b} = \int_{-\alpha_1}^{\alpha_2} \frac{3\{1+cos\alpha_1\,[(\alpha_2-\alpha)\,sin\alpha - cos\alpha]\}^2\,(\alpha_2-\alpha)\,cos\alpha}{2EL\,[sin\alpha + (\alpha_2-\alpha)\,cos\alpha]^3} d\alpha. \tag{2}$$

$$\frac{1}{k_s} = \int_{-\alpha_1}^{\alpha_2} \frac{1.2\,(1+\nu)\,(\alpha_2-\alpha)\,cos\alpha\,(cos\alpha_1)^2}{EL\,[sin\alpha + (\alpha_2-\alpha)\,cos\alpha]} d\alpha. \tag{3}$$

$$\frac{1}{k_a} = \int_{-\alpha_1}^{\alpha_2} \frac{(\alpha_2-\alpha)\,cos\alpha\,(sin\alpha_1)^2}{2EL\,[sin\alpha + (\alpha_2-\alpha)\,cos\alpha]} d\alpha. \tag{4}$$

where E is the elastic modulus of rigidity for gear material, L is the width of tooth and α_1 and α_2 are the angles shown in Fig. 1.

The gear fillet-foundation stiffness (k_f) is calculated as per equation given below [19].

$$\frac{1}{k_f} = \frac{(cos\alpha_m)^2\left\{L^*\left(u_f/s_f\right)^2 + M^*\left(u_f/s_f\right) + P^*\left(1+Q^*\,(tan\alpha_m)^2\right)\right\}}{EL}. \tag{5}$$

where, α_m is acting pressure angle of gear tooth and 'L' is the width of tooth. The coefficients L^*, M^*, P^* and Q^* are computed using Eq. (6)

$$X_i^*\left(h_{fi}, \theta_f\right) = \frac{A_i}{\theta_f^2} + B_i h_{fi}^2 + \frac{C_i h_{fi}}{\theta_f} + \frac{D_i}{\theta_f} + E_i h_{fi} + F_i. \tag{6}$$

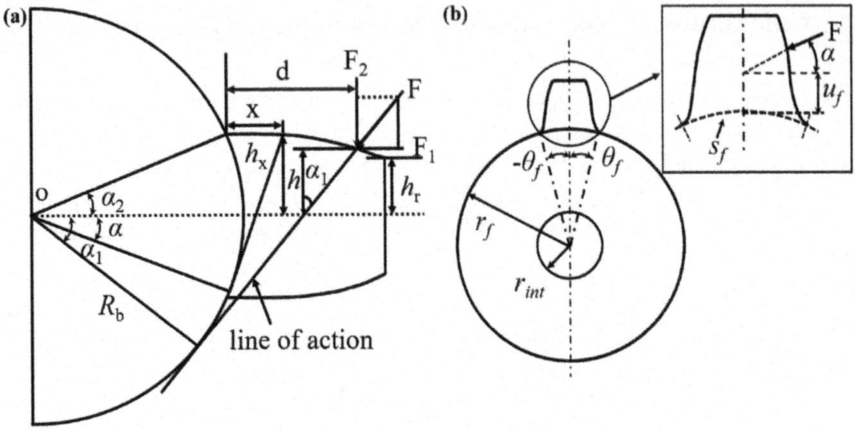

Fig. 1 **a** Healthy tooth model, **b** Geometrical parameter of the fillet-foundation deflection

Table 1 Values of coefficient used in Eq. (6) [20]

	$A_i * 10^5$	$B_i * 10^3$	$C_i * 10^4$	$D_i * 10^3$	$E_i * 10^2$	F_i
$L^* (h_{fi},\theta_f)$	−5.574	1.9986	−2.3015	4.7702	2.71	6.8045
$M^* (h_{fi},\theta_f)$	60.111	28.100	−83.431	−9.9256	16.24	0.9086
$P^* (h_{fi},\theta_f)$	−50.952	185.50	0.0538	53.3	28.95	0.9236
$Q^* (h_{fi},\theta_f)$	−6.2042	9.0889	−4.0964	7.8297	−14.72	0.6904

where, X_i^* denotes the coefficients L*, M*, P* and Q*. The other terms such as h_f, r_{int}, r_f and $h_{fi}= r_f/r_{int}$, are the geometric parameter as described in Fig. 1b, the values of coefficient used in Eq. (6) are given in Table 1.

Thus, the total TVMS for a single-tooth-pair in contact can be expressed as

$$K_m = \frac{1}{\frac{1}{k_h} + \frac{1}{k_{b1}} + \frac{1}{k_{s1}} + \frac{1}{k_{a1}} + \frac{1}{k_{f1}} + \frac{1}{k_{b2}} + \frac{1}{k_{s2}} + \frac{1}{k_{a2}} + \frac{1}{k_{f2}}}. \tag{7}$$

where, k_a, k_b, k_s, k_h and k_f represent the axial compressive, bending, shear, Hertzian stiffness and fillet foundation deflection. Subscript 1 and 2, used in the expression, denote the pinion and gear, respectively.

When more than one teeth are in contact then the TVMS is computed using expression

$$K_T = \sum_{i=1}^{n} \frac{1}{\frac{1}{k_h} + \frac{1}{k_{b1i}} + \frac{1}{k_{s1i}} + \frac{1}{k_{a1i}} + \frac{1}{k_{f1i}} + \frac{1}{k_{b2i}} + \frac{1}{k_{s2i}} + \frac{1}{k_{a2i}} + \frac{1}{k_{f2i}}}. \tag{8}$$

where n denotes the number of teeth pair in meshing

In the present study, a contact ratio ($C = 2.1667$) of pinion gear pair is taken into consideration for analysis of single stage simple gear train. For the given contact ratio, three pairs of teeth are in 0.1167For a gear which revolves at a speed of N rpm and has Z numbers of teeth, the gear meshing period (in second) of the gear can be calculated as $t_m = \frac{60}{N*Z}$ and the corresponding gear mesh frequency $f_m = \frac{N*Z}{60}$.

2.2 For Faulty Condition

In this study, the gear fault like a broken tooth (with different percentage of a broken tooth) on the pinion is considered. With the increase of percentage of a broken tooth (Fig. 2), the corresponding contact with the mating gear is going to decrease and becomes out of contact when it is completely broken. Therefore, the mesh stiffness corresponding to this tooth will affect the time-varying mesh stiffness. If the ith tooth on the pinion is broken, the corresponding gear mesh stiffness is going to be $x * K_i(t)$ (x is the percentage of a broken tooth). If the ith tooth on the pinion is completely broken, the corresponding gear mesh stiffness $K_i(t)$ is zero. Therefore, when a broken tooth comes in the contact zone, the time-varying stiffness can be expressed as

$$K_m(t) = \begin{cases} K_{i-2}(t) + K_{i-1}(t) + K_i(t) \ \forall \ (i)\, t_m \leq t < i\,(C-2)\, t_m \\ K_{i-1}(t) + K_i(t) \quad \forall \ (i)\,(C-2)\, t_m \leq t < (i+1)\, t_m \\ K_{i-1}(t) + K_i(t) + K_{i+1}(t) \ \forall \ (i+1)\, t_m \leq t < (i+1)\,(C-2)\, t_m \\ K_i(t) + K_{i+1}(t) \quad \forall \ (i+2)\, t_m \leq t < (i+2)\,(C-2)\, t_m \\ K_i(t) + K_{i+1}(t) + K_{i+2}(t) \ \forall \ (i+2)\, t_m \leq t < (i+2)\,(C-2)\, t_m \end{cases} \tag{9}$$

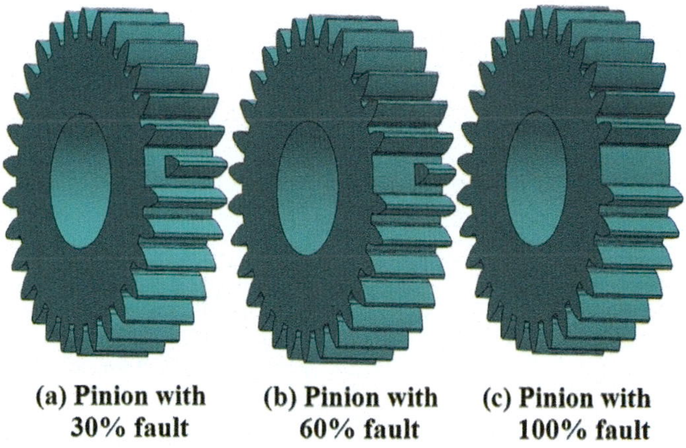

Fig. 2 Pinion with increasing level of fault

Expressions (1–9) are used to determine the time-varying mesh stiffness for a high contact ratio of healthy and faulty spur gear and the corresponding TVMS is then used in dynamic simulation of 9-DOF spur gear system to find the response under different operating speed.

3 Modelling of 9-DOF Spur Gear System

A simplified 9-DOF simple single-stage spur gear system [21] is modelled in MATLAB-Simulink environment to find the response of the system with and without faulty tooth. In modelling the system, it is assumed that the gear and the pinion of the system are connected to the load and driver through different shafts. The effects due to coupling (connection between driven and driven shaft) is neglected in the model. The accelerometer used for measuring the response is mounted on a plate. This model has 5-translational DOF (one for each pinion, gear, bearing foundation and plate) and 4-rotational DOF (one DOF for each pinion, gear, load and driver). All the important parts (pinion, gear, bearing foundation and plate) are considered to be a spring-mass-damper system in modelling (Fig. 3). The equations of motion for the given system [21] is derived as

$$I_d\ddot{\theta}_d + c_{t1}(\dot{\theta}_d - \dot{\theta}_1) + k_{t1}(\theta_d - \theta_1) = T_d, \tag{10}$$

$$I_1\ddot{\theta}_1 + c_{t1}(\dot{\theta}_1 - \dot{\theta}_d) + k_{t1}(\theta_1 - \theta_d) = -NR_{b1}, \tag{11}$$

$$I_2\ddot{\theta}_2 + c_{t2}(\dot{\theta}_2 - \dot{\theta}_l) + k_{t2}(\theta_2 - \theta_l) = NR_{b2}, \tag{12}$$

$$I_l\ddot{\theta}_l + c_{t2}(\dot{\theta}_l - \dot{\theta}_2) + k_{t2}(\theta_l - \theta_2) = -T_l, \tag{13}$$

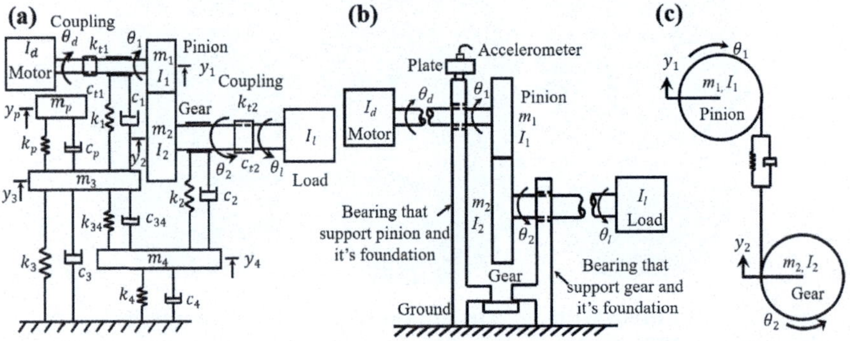

Fig. 3 Spur gear system with nine degrees of freedom **a** spring mass damper system, **b** schemetic and **c** gear interaction

$$m_1\ddot{y}_1 + c_1(\dot{y}_1 - \dot{y}_3) + k_1(y_1 - y_3) = N, \tag{14}$$

$$m_2\ddot{y}_2 + c_2(\dot{y}_2 - \dot{y}_4) + k_2(y_2 - y_4) = -N, \tag{15}$$

$$\begin{aligned} m_3\ddot{y}_3 + c_{34}(\dot{y}_3 - \dot{y}_4) + k_{34}(y_3 - y_4) + c_1(\dot{y}_3 - \dot{y}_1) + k_1(y_3 - y_1) + \\ c_p(\dot{y}_3 - \dot{y}_p) + k_p(y_3 - y_p) + c_3\dot{y}_3 + k_3y_3 = 0, \end{aligned} \tag{16}$$

$$\begin{aligned} m_4\ddot{y}_4 + c_{34}(\dot{y}_4 - \dot{y}_3) + k_{34}(y_4 - y_3) + c_2(\dot{y}_4 - \dot{y}_2) + \\ k_2(y_4 - y_2) + c_4(\dot{y}_4) + k_4y_4 = 0, \end{aligned} \tag{17}$$

$$m_p\ddot{y}_p + c_p(\dot{y}_p - \dot{y}_3) + k_p(y_p - y_3) = 0. \tag{18}$$

where N is the dynamic meshing force given by

$$N = k_m(R_{b1}\theta_1 - R_{b2}\theta_2 - y_1 + y_2) + c_m(R_{b1}\dot{\theta}_1 - R_{b2}\dot{\theta}_2 - \dot{y}_1 + \dot{y}_2). \tag{19}$$

In the expression terms, I_1, I_2, I_d and I_l represent inertia of pinion, gear, drive and load and m_p, m_1 and m_2 represent the mass of plate, pinion and gear. The mass of foundation with bearing for pinion and gear is represented by m_3 and m_4, respectively. The transverse stiffness of each shaft is represented by k_1 and k_2, and the transverse damping by c_1 and c_2. A spring with stiffness k_{34} and a damper with dampening c_{34} join the two bearing foundations. The primary base is joined to the pinion and gear bearing foundations, respectively, by springs with stiffness k_3, k_4 and dampers with damping c_3, c_4. The structural stiffness and damping between the plate and the bearing that supports the pinion are represented by c_p and k_p, respectively. k_{t1} and k_{t2} are torsional stiffness and c_{t1} and c_{t2} torsional damping between the shafts and the load and the shaft and the drive. In this model, the transverse vibration of the gear is taken along the gear line of action as shown in Fig. 3.

3.1 Modelling of Higher Contact Ratio (H.C.R) Spur Gear System

There are number of ways to construct a spur gear system with high contact ratio. For an addendum of 1.35 (as shown in Fig. 4) and normal pressure angle of 20°, high contact ratio (2.1667) spur gear system can be modelled. Proper geometrical parameters of gear and pinion should be taken to avoid interference between the matting gears and banding failure of the tooth. The specification of high contact ratio spur gear model is listed in Table 2.

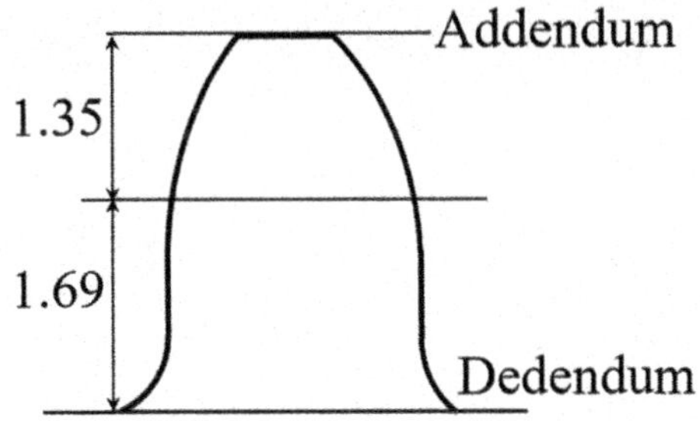

Fig. 4 1.35 addendum spur gear tooth

Table 2 Specification of H.C.R. spur gear system

Parameter	Pinion	Gear
Number of teeth	30	36
Mass moment of inertia (kg.m2)	$2 * 10^{-4}$	$2 * 10^{-3}$
Mass (kg)	0.4439	0.6393
Speed (rpm)	600	500
Pressure angle (in deg.)	20	
Module (mm)	2	
Addendum coefficient (f_a)	1.35	
Dedendum coefficient (f_d)	1.69	
Tooth width (mm)	15	

3.2 Simulation Setup

Equations 10–18 are solved in MATLAB-Simulink environment using the ODE 45 solver with a relative tolerance of $1 * 10^{-4}$. Table 3 lists the parameters of nine-DOF simple spur gear systems. The MATLAB-Simulink model of the corresponding gear system is shown in Fig. 5. A MATLAB function block is used to calculate the dynamic forces N and the corresponding moment M_i acting on the dynamic system. Calculation of TVMS for both healthy and faulty conditions is included in the code of the MATLAB function. There are nine inputs given to this block diagram in the MATLAB-SIMULINK model such as linear displacement (y_1, y_2), angular displacement (θ_1, θ_2), linear velocity ($\dot{y}_1$, $\dot{y}_2$), and angular velocity ($\dot{\theta}_1$, $\dot{\theta}_2$) and the simulation time (t).

Table 3 Inertia, stiffness and damping parameters of 9-DOF model

Inertia parameter	Stiffness parameter	Damping parameter
$m_1 = 0.4439$ (kg)	$k_1 = 2.5 * 10^7$ (N/m)	$c_1 = 43.5$ (N.s/m)
$m_2 = 0.6393$ (kg)	$k_2 = 2.5 * 10^7$ (N/m)	$c_2 = 54.37$ (N.s/m)
$m_3 = 0.33$ (kg)	$k_3 = 7.8 * 10^8$ (N/m)	$c_3 = 321$ (N.s/m)
$m_4 = 0.55$ (kg)	$k_4 = 7.8 * 10^8$ (N/m)	$c_4 = 414$ (N.s/m)
$m_p = 0.23$ (kg)	$k_{t1} = 276$ (N/m)	$c_{34} = 297$ (N.s/m)
$I_1 = 2 * 10^{-4}$ (kg.m^2)	$k_{t_2} = 248$ (N/m)	$c_{t_1} = 0.0459$ (N.s/m)
$I_2 = 4.14 * 10^{-4}$ (kg.m^2)	$k_{34} = 10.7 * 10^8$ (N/m)	$c_{t_2} = 0.0611$ (N.s/m)
$I_d = I_l = 4.1 * 10^{-3}$ (kg.m^2)	$K_p = 1 * 10^8$ (N/m)	$c_p = 74$ (N.s/m)

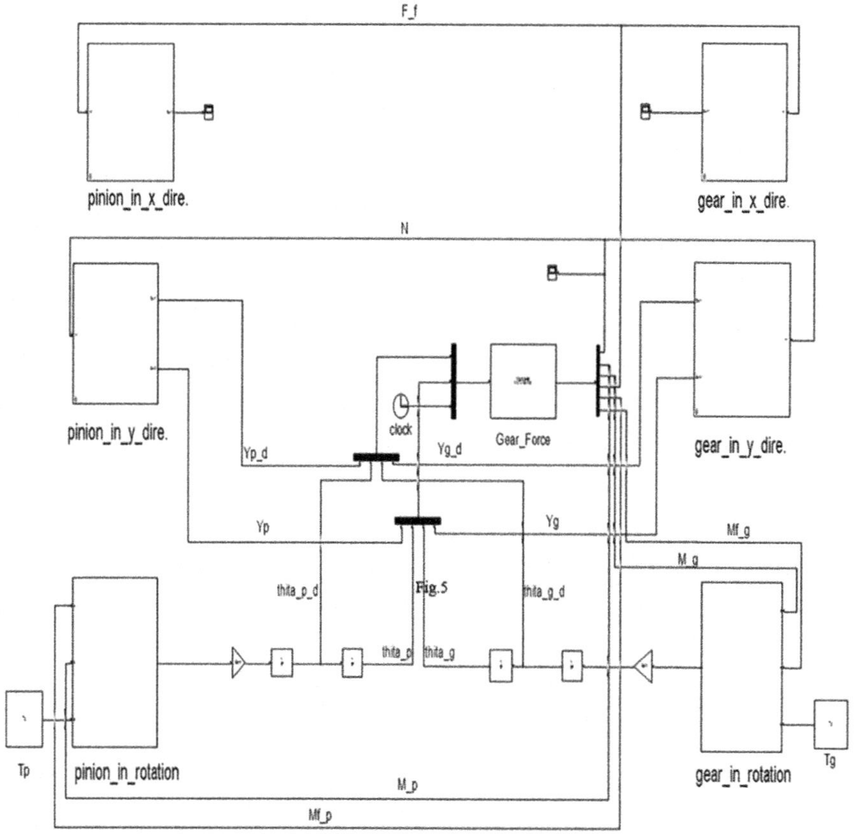

Fig. 5 MATLAB-Simulink model for 9-DOF spur gear system

4 Result and Discussion

The dynamic models of healthy as well as faulty gear system as discussed in the previous section are used to study the response of vibration under constant speed and constant load conditions. An analytical method is used to calculate the TVMS of spur gear pair and is incorporated into the dynamic equations of motion of a 9-DOF system.

A constant driving torque (T_d) of 20 Nm is applied to the system and the load torque is kept constant and set at (T_d*S.R), where, S.R is the speed ratio. The analytical result of the TVMS of the healthy gear pair is shown in Fig. 6. High contact ratio spur gear pair is considered for the analysis, so during engagement of gear teeth, double and triple teeth are engaged alternately. In Fig. 6, the term $C * t_m$ denotes the time period of an ith tooth from engagement to disengagement. When a tooth is broken, the contact area between the tooth is also affected. In case of faulty simple spur gear system, TVMS varies with faults. Figure 7 shows a comparative study of TVMS with and without faults in case of a simple spur gear system, and it shows that with fault in gear tooth the mesh stiffness varies significantly. With the increase in percentage of broken tooth, there is a linear variation in time period for TVMS deviation. For example, for a 30% of broken tooth in one tooth of a pinion, the stiffness deviates for a time period of $0.3 * C * t_m$. Similarly, for 60% and 100% broken tooth the TVMS deviate for a time period of $0.6 * C * t_m$ and $C * t_m$, respectively. Therefore, in case of healthy system, the deviation time is zero and it increases up to $C * t_m$ for a completely broken tooth.

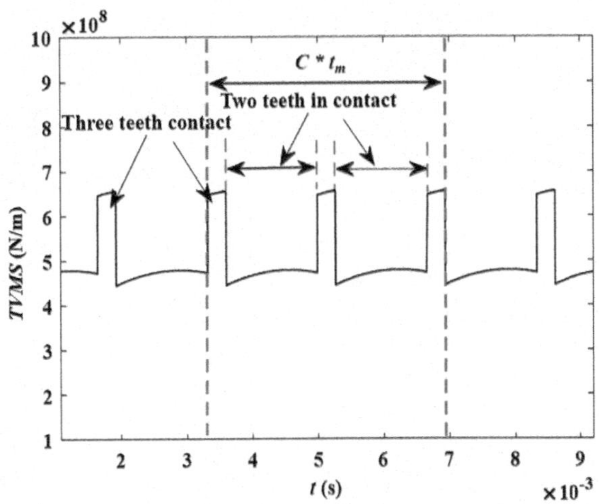

Fig. 6 TVMS of healthy spur gear system

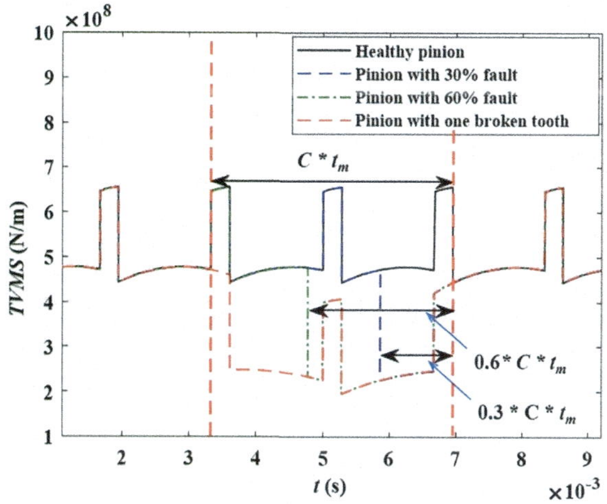

Fig. 7 Comparison of TVMS of healthy and faulty condition spur gear system

4.1 Time-Domain Analysis

One of the major factors for excitation in case of a gear system is the equivalent stiffness that varies with the movement of the contact point along the line of action. To construct a precise gear transmission system model that can completely reflect the dynamic response characteristics of gear teeth, it is crucial to design an appropriate method for determining the equivalent stiffness of the gear teeth. In the initial phase of gear failure, the fault characteristics are smaller due to which it is difficult to identify; hence, a preventive fault diagnosis is always recommended.

The TVMS of gear with different level of broken tooth is introduced into the dynamic equations of motion (Eqs. 10 to 18). These equations of motions are solved using the Runge-Kutta method and the corresponding vibration responses are monitored. In this analysis, pinion rotates at a constant rotational frequency of $f_p = 10$ Hz $\left(t_{rp} = 0.06sec\right)$. Thus, the corresponding gear frequency and mesh frequency are $f_g = 8.333$ Hz and $f_m = 300$ Hz. In case of a broken tooth, the characteristic frequency of the broken tooth is same as the rotational frequency of the pinion shaft. Figures 8 and 9 illustrate the vibration acceleration response of pinion and gear in the time domain for a short time window of 0.07 s. The symbol A_p and A_g represents the amplitude of acceleration of pinion and gear. The response of pinion and gear in the y-direction of the healthy spur gear system is shown in Figs. 8a and 9a. The amplitude of each mesh period is the same for both the pinion and gear which shows that there is no irregularities in the amplitude of vibration response. Figures 8 and 9, (b) to (d) indicate the response of pinion and gear in three different levels of a broken tooth (i.e. 30%, 60% and one complete broken tooth on pinion). The red square box in those figures indicates the location of the broken tooth. It can be further inferred that peaks of higher amplitudes are present in the vibration acceleration response as compared to the healthy one. These peaks are observed due to the presence of

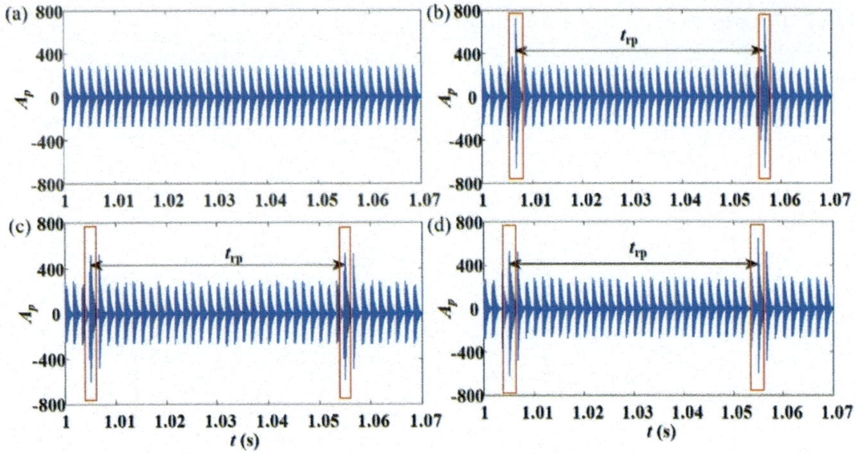

Fig. 8 Vibration response of pinion for a spur gear system with pinion, **a** healthy **b** a tooth 30% broken, **c** a tooth 60% broken and **d** a tooth completely broken

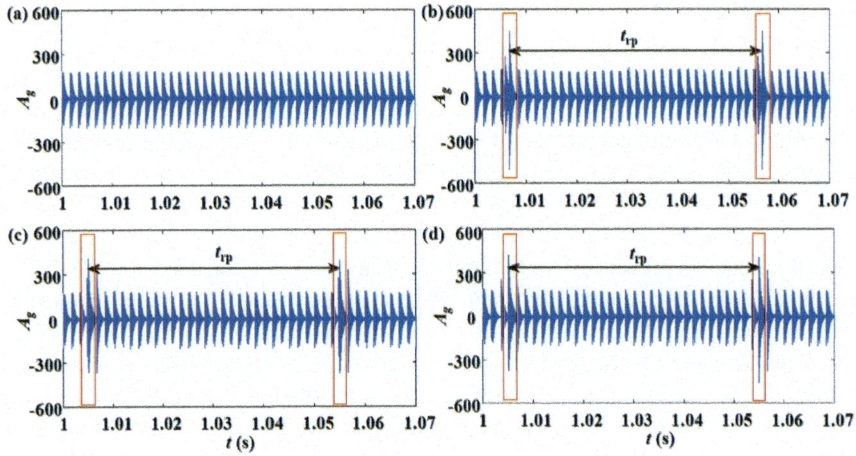

Fig. 9 Vibration response of gear for a spur gear system with pinion, **a** healthy, **b** a tooth 30% broken, **c** a tooth 60% broken and **d** a tooth completely broken

different levels of missing tooth in the meshing zone, which induces a larger amount of impulsive dynamic force in the system. The span between peaks is the pinion rotation period $\left(t_{rp}\right)$, as the missing tooth is on the pinion and the span between the consecutive peaks gives the time period for the loss and gain of contact of the teeth.

4.2 Frequency Domain Analysis

Envelope analysis is used as the post-processing step for getting the denoised signal. The envelope spectrum of the corresponding signal (Figs. 8 and 9) is shown in Figs. 10 and 11. The symbols $A_p^e\ (j\omega)$ and $A_g^e\ (j\omega)$ are the amplitude of envelope spectrum of the corresponding signal. Figures 10 and 11 (a) show the envelop spectrum of healthy pinion and gear and (b) to (d) of the respective figures show the envelope spectrum of pinion and gear with different levels of broken tooth (i.e. 30%, 60% and one complete broken tooth on pinion).

The envelope spectrum of pinion and gear in healthy conditions is characterized by gear mesh frequency (f_m) and its harmonics (n*f_m). In the envelope spectrum of a broken tooth, amplitude modulation occurs on both pinion and gear due to the presence of high-magnitude peaks in the vibration signal. Because of the amplitude modulation, higher magnitude sidebands are formed around gear mesh frequencies and their harmonics due to the presence of faults. The sidebands are characterized as (i*$f_m \pm$ j*f_{rp}). In the expression, i, j are integers and f_{rp} is the pinion rotational frequency. Characterization of sidebands with f_{rp} indicates the source of the fault in the pinion. It can be inferred from the envelope spectrum of the signal that with the increase in the level of faults, magnitude of peaks at gear mesh frequency and its harmonics increase significantly and the amplitude of its sidebands also increases significantly. These results validate the earlier reported work by Kumar et al. [21] in the analysis of bevel gear train with increasing level of fault.

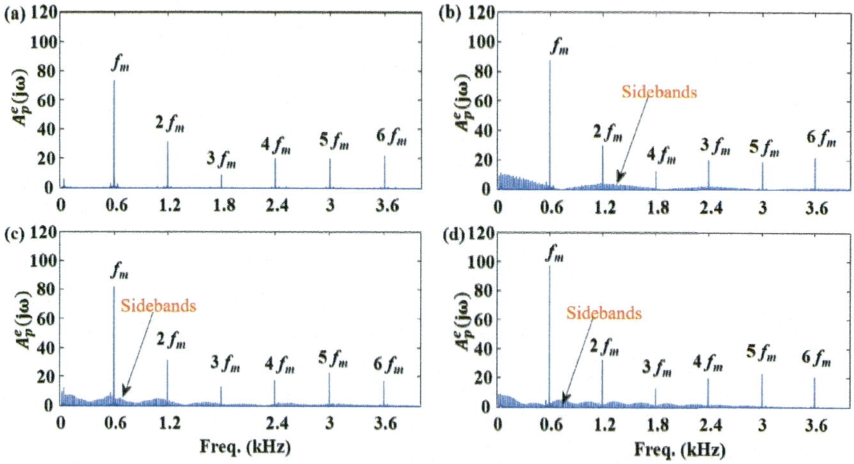

Fig. 10 Envelope spectrum of pinion for a spur gear system with pinion, **a** healthy **b** a tooth 30% broken, **c** a tooth 60% broken and **d** a tooth completely broken

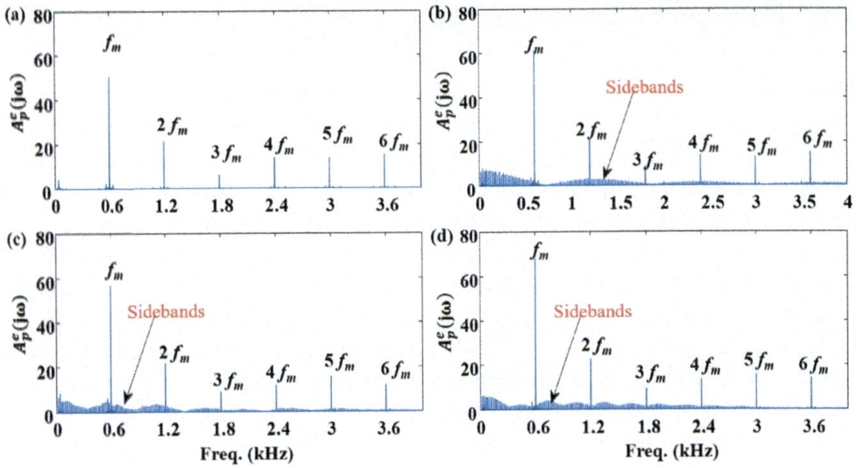

Fig. 11 Envelope spectrum of gear for a spur gear system with pinion, **a** healthy, **b** a tooth 30% broken, **c** a tooth 60% broken and **d** a tooth completely broken

5 Conclusions and Contributions

The conclusions and contributions of the present work can be summarized as follows:

- A dynamic model of 9-DOF spur gear system is developed in MATLAB-Simulink environment.
- A gear fault, with increasing order of a broken tooth (i.e. 30%, 60% and one complete broken tooth) is considered for this analysis.
- When a broken tooth of pinion makes contact with tooth of gear during meshing, the mesh stiffness is reduced significantly, and the mesh stiffness reduces further with the increase in the level of a broken tooth.
- Time and frequency domain analysis is used to study the response of the gear box with and without fault.
- In time domain, it can be observed that, with the increase in the level of broken tooth, there is an increase in the magnitudes of peaks which indicates the impact of induced impulsive force generated due to the presence of tooth broken in the system.
- In the envelope spectrum, it can be further observed that with the increase in percentage of a broken tooth there is an increase in the amplitude of gear mesh frequency and its harmonics and the number of sidebands.

References

1. Haword I, Jia S, Wang J (2001) The dynamic modeling of a spur gear in mesh including friction and a crack. Mech Syst Signal Process 15:831–853
2. Chen Z, Shao Y (2011) Dynamic simulation of spur gear with tooth root crack propagating along tooth width and crack depth. Eng Fail Anal 18:2149–2164
3. Saxena A, Parey A, Chouksey M (2016) Time varying mesh stiffness calculation of spur gear pair considering sliding friction and spalling defects. Eng Fail Anal 70:200–211
4. Saxena A, Chouksey M, Parey A (2017) Effect of mesh stiffness of healthy and cracked gear tooth on model and frequency response characteristics of geared rotor system. Mech Mach Theory 107:261–273
5. Chen Z, Zhai W, Shao Y, Wang K, Sun G (2016) Analytical model for mesh stiffness calculation of spur gear pair with non-uniformly distributed tooth root crack. Eng Fail Anal 66:502–514
6. Yang DCH, Lin JY (1987) Hertzian damping, tooth friction and bending elasticity in gear impact dynamics. J Mech Transm Autom Des 109:189–196
7. Tian X (2004) Dynamic simulation for system response of gearbox including localized gear faults. M.S. thesis, Edmonton, University of Alberta, Canada
8. Wu S, Zuo MJ, Parey A (2008) Simulation of spur gear dynamics and estimation of fault growth. J Sound Vib 317:608–624
9. Chen Z, Shao Y (2013) Mesh stiffness calculation of a spur gear pair with tooth profile modification and tooth root crack. Mech Mach Theory 62:63–74
10. Pandya Y, Parey A (2013) Crack behavior in a high contact ratio spur gear tooth and its effect on mesh stiffness. Eng Fail Anal 34:69–78
11. Mohammed OD, Rantatalo M, Aidanpaa JO (2015) Dynamic modelling of a one-stage spur gear system and vibration-based tooth crack detection analysis. Mech Syst Signal Process 54–55:293–305
12. Kumar R, Mitra RK, Dewangan R, Mishra C (2023) Modelling and diagnosis of faults in simple bevel gear train. Wear 524–525, 204881 (1–15)
13. Staszewski WJ, Tomlinson GR (1994) Application of the wavelet transform to fault detection in a spur gear. Mech Syst Signal Process 8:289–307
14. Zheng H, Li Z, Chen X (2002) Gear fault diagnosis based on continuous wavelet transform. Mech Syst Signal Process 16:447–457
15. Rajeswari C, Sathiyabhama B, Devendiran S, Manivannan K (2014) A gear fault identification using wavelet, rough set based GA, ANN and C4.5 algorithm. Proced Eng 97:1831–1841
16. Li B, Zhang X, Wu J (2017) New Procedure for gear fault detection and diagnosis using instantaneous angular speed. Mech Syst Signal Process 85:45–428
17. Igba J, Alemzadeh K, Durugbo C, Eiriksson ET (2016) Analysing RMS and peak values of vibration signals for conditional monitoring of wind turbine gearboxes. Renew Energy 91:90–106
18. Wang W, Wong AK (2002) Autoregressive model-based gear fault diagnosis. J Vib Acoust 124:172–179
19. Sainsot P, Velex P, Duverger O (2004) Contribution of gear body to tooth deflections-a new bidimensional analytical formula. J Mech Des 126:748–752
20. Chaari F, Baccar W, Abbes MS, Haddar M (2008) Effect of spalling or tooth breakage on gearmesh stiffness and dynamic response of a one-stage spur gear transmission. Eur J Mech A/Solids 27:691–705
21. Omar FK, Moustaf KAF, Emam S (2011) Mathematical modeling of gearbox including defects with experimental verification. J Vib Control 18:1310–1321

The manufacturer's authorised representative in the EU is Springer Nature Customer Service Centre GmbH, Europaplatz 3, 69115 Heidelberg, Germany. If you have any concerns regarding our products, please contact ProductSafety@springernature.com

Printed and bound by CPI Group (UK) Ltd, Croydon, CR0 4YY
01/10/2026
02225985-0004